Springer Theses

Recognizing Outstanding Ph.D. Research

Aims and Scope

The series "Springer Theses" brings together a selection of the very best Ph.D. theses from around the world and across the physical sciences. Nominated and endorsed by two recognized specialists, each published volume has been selected for its scientific excellence and the high impact of its contents for the pertinent field of research. For greater accessibility to non-specialists, the published versions include an extended introduction, as well as a foreword by the student's supervisor explaining the special relevance of the work for the field. As a whole, the series will provide a valuable resource both for newcomers to the research fields described, and for other scientists seeking detailed background information on special questions. Finally, it provides an accredited documentation of the valuable contributions made by today's younger generation of scientists.

Theses are accepted into the series by invited nomination only and must fulfill all of the following criteria

- They must be written in good English.
- The topic should fall within the confines of Chemistry, Physics, Earth Sciences, Engineering and related interdisciplinary fields such as Materials, Nanoscience, Chemical Engineering, Complex Systems and Biophysics.
- The work reported in the thesis must represent a significant scientific advance.
- If the thesis includes previously published material, permission to reproduce this must be gained from the respective copyright holder.
- They must have been examined and passed during the 12 months prior to nomination.
- Each thesis should include a foreword by the supervisor outlining the significance of its content.
- The theses should have a clearly defined structure including an introduction accessible to scientists not expert in that particular field.

More information about this series at http://www.springer.com/series/8790

Takahiro Nemoto

Phenomenological Structure for the Large Deviation Principle in Time-Series Statistics

A Method to Control the Rare Events in Non-equilibrium Systems

Doctoral Thesis accepted by
Kyoto University, Kyoto, Japan

Springer

Author
Dr. Takahiro Nemoto
Department of Physics
Kyoto University
Kyoto
Japan

Supervisor
Prof. Shin-ichi Sasa
Department of Physics
Kyoto University
Kyoto
Japan

ISSN 2190-5053 ISSN 2190-5061 (electronic)
Springer Theses
ISBN 978-981-10-1318-8 ISBN 978-981-287-811-3 (eBook)
DOI 10.1007/978-981-287-811-3

Springer Singapore Heidelberg New York Dordrecht London

Softcover re-print of the Hardcover 1st edition 2016

Printed on acid-free paper

Springer Science+Business Media Singapore Pte Ltd. is part of Springer Science+Business Media (www.springer.com)

Parts of this thesis have been published in the following journal articles:

- T. Nemoto and S. Sasa, Phys. Rev. Lett. **112**, 090602 (2014),
- T. Nemoto, V. Lecomte, S. Sasa and F. van Wijland, J. Stat. Mech. (2014) P10001,
- T. Nemoto, Phys. Rev. E **85**, 061124 (2012).

Supervisor's Foreword

A large deviation function characterises the asymptotic behaviour of the probability of the average value for stochastic variables. The theory for large deviation functions was developed as a branch of probability theory about 50 years ago. Despite the progress in mathematics, many physicists are not so familiar with this large deviation theory. However, the situation is now changing. For a wide class of systems out of equilibrium, a novel symmetry property, which is now called the fluctuation theorem, was discovered. Such a simple and universal relation in physics was then clearly expressed by using the large deviation function of time-averaged entropy production rate. This example suggests that the study of fluctuations on the basis of the large deviation theory can discover new physical laws. Within this context, the Springer Thesis by Takahiro Nemoto provides us a thoughtful insight into the intersection between physics and the large deviation theory.

An important point of view is that the basic concept of large deviation theory was traced back to Einstein's fluctuation theory. On the one hand, large deviation functions are determined from fluctuation properties of thermodynamic extensive variables, and on the other hand, thermodynamic functions are obtained by the measurement of the heat capacity and the equation of state. Einstein's fluctuation theory begins with a hypothesis where these large deviation functions and thermodynamic functions are equivalent up to a multiplicative constant. This hypothesis could be proved on the principle of equal weight, from which many non-trivial predictions were confirmed experimentally, leading to the establishment of equilibrium statistical mechanics. Takahiro Nemoto seeks a similar framework for time-averaged quantities in systems out of equilibrium, by which large deviation functions can be obtained without measuring fluctuations. He calls such a framework "phenomenological structure for the large deviation principle" as an extension of Einstein's fluctuation theory.

Readers of this Springer Thesis can obtain the most advanced knowledge on the phenomenological structure for the large deviation principle with a background idea, a new theoretical method, novel formulas, and illustrative examples.

In addition to the fundamental aspects of the problem, this framework also involves a proposal of a new method for rare-event sampling, which could be another interesting topic. I am sure that graduate students in theoretical physics as well as researchers can enjoy reading this Springer Thesis.

Kyoto, Japan
September 2015

Prof. Shin-ichi Sasa

Acknowledgments

First I would like to thank Shin-ichi Sasa, who has supervised me for five and a half years, since the final year of my undergraduate course. His research and his passion for science have influenced me greatly. I cannot imagine how I would have managed if he had not been my supervisor.

Next I would like to thank Frédéric van Wijland and Vivien Lecomte, from Paris Diderot, who are the collaborators on the second part (Chap. 3) of this thesis. They gave me the opportunity to work with them in their laboratory many times when I was visiting Paris. I would also like to thank Jorge Kurchan from the École Normale Supérieure and Julien Tailleur from Paris Diderot for the many discussions that we had. We are currently collaborating on some ongoing research. They allowed me to work in their laboratory many times as well.

I would also like to say thank you to the individuals who shared the same laboratory with me: Taiki Haga, Masato Itami, Michikazu Kobayashi, Shuhei Kurita, Yasuhiro Mochizuki, Tomokatsu Onaga, Daisuke Sato, Shigeru Shinomoto, Toshiaki Shintani, Hiroyoshi Takagi, Mitsusuke Tarama, Masahiko Ueda, and Yuzuru Yamanaka. Through various discussions, they helped me many times on my research and also in my problems in daily life.

I thank, as well, the members of the fluid physics group in the same department, with whom I shared the same room from the second year of my Ph.D. course: Shunsuke Akita, Yoshiki Hiruta, Takeshi Matsumoto, Ryo Murakami, Toh Sadayoshi, Kentaro Takagi, Naohiro Temmoku, and Toshiki Teramura. The conversations that I had with them were very motivating and were, for example, on the current hot topics of fluid dynamics research.

During the several national and international conferences that I attended, I was able to meet many excellent researchers. We discussed my research and they were able to give me many useful comments on it. I would like to thank them: Florian Angeletti, Takeaki Araki, Thierry Bodineau, Freddy Bouchet, Raphaël Chetrite, Krzysztof Gawędzki, Hisao Hayakawa, Robert Jack, Christian Maes, Karel Netocný, Kunimasa Miyazaki, Tomoyuki Obuchi, Jun Ohkubo, Hiroki Ohta, Masayuki Ohzeki, Takahiro Sagawa, Keiji Saito, Ken Sekimoto, Peter Sollich,

Shinji Takesue, Kazumasa Takeuchi, Hal Tasaki, Cristina Toninelli, Toshiyuki Tanaka, and Hugo Touchette.

I would like to express my thanks to the students who are also graduating this year: Tommaso Brotto, Tatsuhiko Ikeda, Sosuke Ito, Kyogo Kawaguchi, Kiyoshi Kanazawa, Tanguy Laffargue, Yohei Nakayama, Alexandre Solon, and Sho Sugiura. The discussions with them were very stimulating and helped me to advance in my work.

Finally, I want to thank my family for supporting me during my studies, and I also thank my fiancée, Jeanne, for cheering me on during my final year of Ph.D. work (even for the interruptions!) and for reading this manuscript many times.

Contents

Chapter 1
Phenomenological Structure for the Large Deviation Principle

1.1 Statistical Mechanics

The motion of moving balls is described by Newton's equation, while the flow of water is described by Navier–Stokes equation. Physics laws take different forms depending on the phenomena in which we are interested. These descriptions are correct in their predictions, although some of them have overlaps in the range of their validity. The air surrounding us is mainly constituted of oxygen and nitrogen molecules, which indicates that the system is described by Newton's law with modeling these molecules as small balls. At the same time, it is a well-known fact that the static property of the ideal gas is summarised by an equation known as the ideal gas law. The former description is legitimate, and the latter one is experimentally confirmed. But, how do we connect these two descriptions? If we restrict ourselves in *equilibrium systems*, a stiff framework connecting microscopic descriptions to macroscopic ones has already been found more than one century ago. It is *the statistical mechanics* which plays this role, and more precisely, which connects microscopic descriptions, namely ones with molecules from Newton's equation or quantum Schrödinger's equation, to *thermodynamics* [1].

The framework of statistical mechanics is useful, therefore it is worth to study the theoretical structure behind it. For the concreteness of the argument, without loss of generality, we restrict ourselves to the ideal gas attached to a reservoir of temperature T (or inverse temperature β). In principle, by describing the system and the reservoir with Newton's equation, we completely specify the behaviour of the system once all of the initial conditions are given. However, in reality such a complete prediction is impossible, so we need to rely on coarse graining methods. Now we recall *the principle of equal a priori probabilities*, which is regarded as one of the basis of statistical mechanics. This principle claims that the equal probability among any state of an isolated system gives a good description of the probability distribution function of the microscopic states. We apply this principle to the whole system by knowing that it is isolated. After tracing out the degrees of freedom in the

T. Nemoto, *Phenomenological Structure for the Large Deviation Principle in Time-Series Statistics*, Springer Theses,
DOI 10.1007/978-981-287-811-3_1

reservoir, we obtain the probability distribution function of the system. By denoting the configuration of the system by Γ, the Hamiltonian of the system by $H(\Gamma)$, and its distribution function by $p_\beta(\Gamma)$, this distribution function is written as

$$p_\beta(\Gamma) \propto e^{-\beta H(\Gamma)}. \tag{1.1}$$

This form of the distribution function is called *canonical distribution function.* We recall that we are now considering the ideal gas, which means that the Hamiltonian of the system has the simple form of $H(\Gamma) = \sum_i \boldsymbol{p}_i^2/(2m)$, where m is the mass of the particle and $\boldsymbol{p}_i$ is the momentum of the ith particle. We thus can calculate any expected value of the macroscopic property, for example, the energy, the pressure and so on. The obtained result is equivalent to the phenomenology confirmed by experiments, such as the ideal gas law. Furthermore, by noticing that the expected value of the macroscopic variables is connected to the derivative of *thermodynamic functions* within the framework of thermodynamics, we also have the formula to obtain the thermodynamic functions. For example, the Helmholtz free energy density f_β is calculated as

$$f_\beta = -\frac{1}{N\beta} \log \sum_\Gamma e^{-\beta H(\Gamma)}, \tag{1.2}$$

where $\sum_\Gamma e^{-\beta H(\Gamma)} \equiv Z_\beta$ is a partition function.

The thermodynamic property of the system is connected to the microscopic description through the formulas (1.1) or (1.2). Furthermore, beyond the thermodynamic properties, these formulas give us another important information. That is rare fluctuations of the system, which is summarised as the theory of *large deviation principle* [2, 3]. We will explain this connection in the next subsection.

1.2 Large Deviation Functions and Thermodynamic Functions

The easiest example to explain the large deviation principle may be the coin-flipping problem. In this problem N coins are flipped. There are no correlations among each flip, therefore the event can be described as *the independent and identically distributed random variables* (iid). We denote by n the number of the coins taking the head-side. The probability of n is then calculated as

$$p(n) = \frac{n!}{N!(N-n)!}\left(\frac{1}{2}\right)^n. \tag{1.3}$$

We define the ratio of the head-side coins x as $x \equiv n/N$. The probability of x is also calculated from (1.3). The numerical example of the probability of x for several N is plotted in the Fig. 1.1. As shown in this figure, the probability has a sharper peak

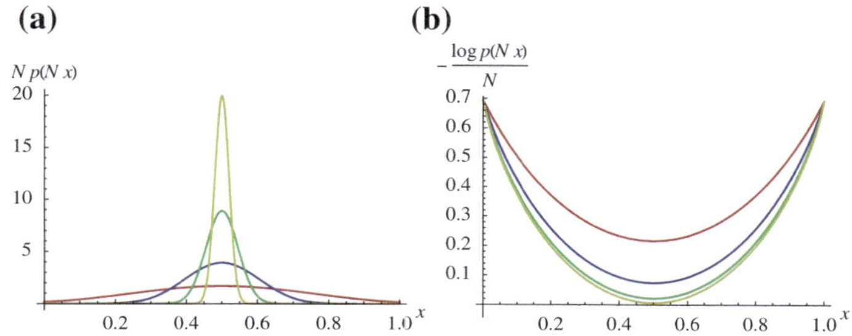

Fig. 1.1 Numerical example of $p(n)$ given as (1.3). **a** shows $p(n)$ and **b** does the logarithm of it. The *red*, *blue*, *green*, and *yellow lines* correspond to $N = 5$, $N = 25$, $N = 125$, and 625, respectively

around $x = 1/2$ with the increasing of N. This represents *the law of large numbers* meaning that the fluctuation of the averaged quantity becomes smaller as the average number N increases. We next consider the speed of the convergence of this law. By applying the Stirling formula to the factorials in (1.3), we obtain the asymptotic form of $p(x)$ as

$$p(x) \sim e^{-NI(x)} \tag{1.4}$$

with

$$I(x) = (1 - x) \log 2(1 - x) + x \log 2x. \tag{1.5}$$

This asymptotic form is called *a large deviation principle*, and the function $I(x)$ is called *a large deviation function* or *a rate function* describing the frequency of rare events. This function takes 0 for typical events and takes values larger than 0 for rare events. It indeed takes 0 for $x = 1/2$ and more than 0 for $x \neq 1/2$ in this example as shown in the right-hand side of Fig. 1.1. The function leads to the probability of the deviation of the law of large numbers, which are quite rare as shown in the left-hand side of Fig. 1.1. This is a simple example of iid, where we can exactly calculate the distribution function. However, the large deviation principle is not restricted to this simple example. In the following chapters, we will analyse Markov dynamics, where the variables are not independent to each other. Even in this case, a large deviation principle is proved [2].

For the case of macroscopic variables, and more precisely, thermodynamic variables in equilibrium systems, the large deviation functions correspond to the thermodynamic functions. We will explain this next. We again consider the same simple example of the ideal gas attached to the reservoir of temperature T. Since the system is attached to the reservoir, the energy is transferred between the system and the reservoir. We then focus on the fluctuation of the energy-density of the system. We denote the density of the number of the state taking the energy U by $\Omega(U) = \int d\Gamma \delta(U - H(\Gamma))$, where $\delta(x)$ is the Dirac delta function. Because the distribution function of Γ is given as (1.1), we evaluate the distribution function of U as

$$p_\beta(U) \propto \Omega(U)e^{-\beta U}. \tag{1.6}$$

Here, we note that one can easily calculate $\Omega(U)$ by exploiting the simplicity of the ideal gas system. We do not write down the explicit form, but just show the asymptotic form of Ω in large system size limit. That is,

$$\Omega(U) \sim e^{Ns(u)}, \tag{1.7}$$

where N is the number of the particles, u is defined as U/N and $s(u)$ is a function that does not depend on N. The asymptotic form (1.7) is generally true, which is not restricted within this ideal gas example. Rather, it is a necessary condition of the validity of thermodynamics tp a given system. One can identify $s(u)$ as the density of thermodynamic entropy by using the relation (1.2). The argument is as follows: We calculate the right-hand side of (1.2) by using the saddle point approximation with (1.7). The result is

$$f_\beta = \min_u \left[u - \beta^{-1}s(u)\right], \tag{1.8}$$

which means that the free energy is the Legendre transformation of $s(u)$. This Legendre transformation is the same form as the thermodynamic relation between a free energy and an entropy.[1] By noticing that the Legendre transformation is an involution (meaning that when it is applied twice to a function g, the obtained function is the same as the original function g), we thus can identify $s(u)$ as the thermodynamic entropy density determined from the thermodynamic free energy f_β. Finally, with the entropy function $s(u)$ and (1.6), the large deviation function of u is written as

$$p_\beta(u) \sim e^{-NI(u)} \tag{1.9}$$

with

$$I(u) = \beta u - s(u) + \text{const.} \tag{1.10}$$

The large deviation function of the energy density is thus written as the sum of the energy and the entropy density. This result is general, not restricted to the large deviation function of the energy density or to this ideal gas system.

1.3 Phenomenological Structure for the Large Deviation Principle

The thermodynamic function is constructed from the expected values of macroscopic variables. On the other hand, the large deviation functions describe the frequency of rare events. In the book "*thermodynamic formalism*" [4] by D. Ruelle, this structure

[1] We set the Boltzmann constant k_B to be 1.

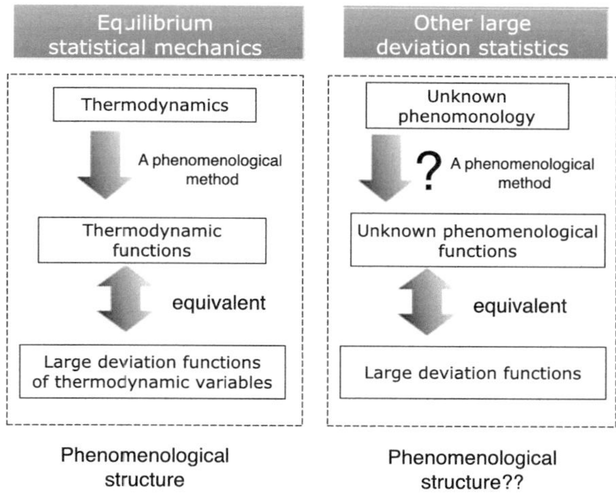

Fig. 1.2 Schematic picture representing phenomenological structure for the large deviation principle

was explicitly mentioned. Due to this book, the formalism of large deviation functions defined in general physical systems, not only in thermodynamical systems, is called *thermodynamic formalism*. There has been a lot of research on thermodynamic formalism, like multi-fractals, chaotic systems and disordered systems. However, there are still less studies directly motivated by this structure. Hereafter, we are going to mention this structure many times, since it is the key of this thesis. For avoiding a lengthy explanation each time, we call this structure *the phenomenological structure for the large deviation principle*, meaning that the large deviation functions can be phenomenologically constructed. In thermodynamics there is this structure, which has been explained in this introduction. (See also Fig. 1.2 for the schematic picture of these relations.) But, *does another statistical system that possesses this structure exist?* This is the question that we study in this thesis, from which some nontrivial features in non-equilibrium physics will be discovered.

For this purpose, we here see the phenomenological structure for the large deviation principle in the equilibrium statistical mechanics with a generally extendable method. The system considered is again the simple example, introduced above, constituted of the ideal gas and the reservoir. To analyse the fluctuation of physical quantities, it is useful to define a cumulant generating function $G(h)$ [2, 3] as

$$G(h) = \lim_{N\to\infty} \frac{1}{N} \log \left\langle e^{hU} \right\rangle . \tag{1.11}$$

Indeed, the large deviation function is connected to $G(h)$ by Legendre transformation:

$$G(h) = \max_{U} \left[hU - I(U) \right] , \tag{1.12}$$

which is equivalent to

$$I(U) = \max_U [hU - G(h)] \tag{1.13}$$

due to the involution property of the Legendre transformation. Equation (1.12) can be easily checked by applying the large deviation form (1.9) in the definition of the cumulant generating function (1.11). These relations represent the equivalence between the cumulant generating function $G(h)$ and the large deviation function $I(U)$.[2] As seen in this discussion, this property is not restricted to equilibrium statistical physics but is also satisfied in general large deviation statistics.

For a phenomenological construction of $G(h)$, we now focus on the derivative of $G(h)$. It becomes

$$\frac{\partial G(h)}{\partial h} = \lim_{N\to\infty} \sum_{\Gamma} \frac{H(\Gamma)}{N} p_\beta(\Gamma; h), \tag{1.14}$$

where we defined *an exponentially biased ensemble* as

$$p_\beta(\Gamma; h) = \frac{p(\Gamma)e^{hH(\Gamma)}}{\sum_\Gamma p(\Gamma)e^{hH(\Gamma)}}. \tag{1.15}$$

This indicates that we can construct $G(h)$ phenomenologically if the exponentially biased ensemble has the corresponding physical system that can be realised in experiments. Indeed, when the biased ensemble has the corresponding physical system, we can replace the right-hand side of (1.14) by the corresponding expected value in the physical system. From the integral of such expected values, the cumulant generating function itself can then be also constructed. For equilibrium statistical physics, the fact that the biased ensemble has the corresponding physical system can be easily checked. Indeed, by utilising the canonical distribution function given as (1.1), we have

$$p_\beta(\Gamma; h) = p_{\beta-h}(\Gamma), \tag{1.16}$$

which means that the exponentially biased ensemble is equivalent to another equilibrium ensemble, whose temperature is modified to $\beta - h$.

In general large deviation statistics, (1.12), (1.13), and (1.14) are satisfied, but (1.16) is not. Since (1.16) is the key for the phenomenological structure for the large deviation principle, *we therefore replace our goal, which is to find another large deviation statistics that possesses the phenomenological structure, to a new goal that is to find another large deviation statistics that has the same property as* (1.16).

---

[2] We should mention that we do not now consider the situation, where phase transitions are taking place. In that case, it may happen that these two functions are not equivalent. See Ref. [3] for the explanation of such examples.

1.4 Large Deviation Statistics of Time-Averaged Quantity in Non-equilibrium System

Recently, large deviation functions of time-averaged quantities in non-equilibrium systems have been gathering attention. The examples are seen in the findings of fluctuation theorem [5–10], an additivity principle for macroscopic currents [11–16], generalised Onsager-Machlup approach [17, 18], dynamical phase transition in kinetically constrained models [19–21], a new Lyapunov function for non-equilibrium steady state [22], exact results on ASEP [23, 24], and some studies related to the phenomenological structure for the large deviation principle in time-series statistics [25–31]. Motivated by these recent results, we here describe a direct benefit of the phenomenological structure in time-series statistics.

1.4.1 Towards a Rare Event Sampling Method in Real Experiments

Let us consider a simple example composed of a Brownian particle with a tilted periodic potential as shown in Fig. 1.3. Most of the time, the Brownian particle will go down the periodic potential because the potential is tilted. However, if we wait for a long time, it is possible to observe the particle climbing up the potential as shown in the right figure of Fig. 1.3. The latter behaviour of the particle is an example of *a rare event*. This rare event is characterised by a large deviation principle as explained next. For the concreteness of the explanation, we describe the particle by using Langevin equation [32]:

$$\frac{dx}{dt} = \frac{1}{\gamma}F(x) + \sqrt{\frac{2T}{\gamma}}\xi(t), \tag{1.17}$$

where x is the position of the particle, $F(x)$ is the tilted periodic potential, γ is a friction constant of the particle, T is the temperature of the system, and $\xi(t)$ is Gaussian white noise that satisfies $\langle\xi(t)\rangle = 0$, $\langle\xi(t)\xi(s)\rangle = \delta(t-s)$. Now, we define a time-averaged velocity of the particle as

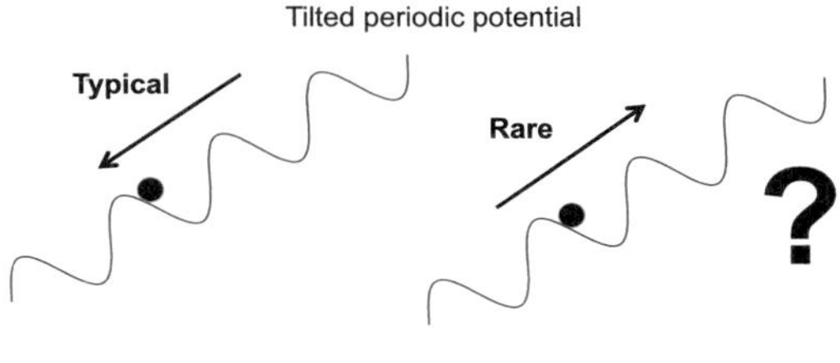

Fig. 1.3 Schematic picture of a Brownian particle on a tilted periodic potential. Most of the time, the particle goes down the potential as shown in the *left* figure, however, the probability of the particle climbing up the potential is not 0

Table 1.1 Correspondence between equilibrium statistical mechanics and time-series statistics

	Statistical mechanics	Time-series statistics
Variables	Configuration of the system Γ	History (or path) of the system $(x(s))_{s=0}^{t}$
Distribution function	Equilibrium distribution	Path probability density $p_F[(x(s))_{s=0}^{t}]$
Large deviation principle	Energy density u $p_\beta(u) \sim e^{-NI(u)}$	Time averaged velocity $V(t)$ $\mathrm{Prob}[V(t) = V] \sim e^{-tI(V)}$
Infinite limit	Infinite system size N limit	Infinite averaging time t limit
Cumulant generating function	$G(h) = \lim_{N\to} \frac{1}{N} \log \langle e^{hNu} \rangle$	$G(h) = \lim_{t\to} \frac{1}{t} \log \langle e^{htV} \rangle$
Exponentially Biased ensemble	$p_\beta(\Gamma; h) \propto p_\beta(\Gamma) e^{hH(\Gamma)}$	$p_F[(x(s))_{s=0}^{t}; h]$ $\propto p_F[(x(s))_{s=0}^{t}] e^{thV(t)}$
Condition for phenomenological structure	$p_\beta(\Gamma; h) = p_{\beta-h}(\Gamma)$	$p_F[(x(s))_{s=0}^{t}; h]$ $= p_{F+\delta w}[(x(s))_{s=0}^{t}]$

$$V(t) = \frac{1}{t} \int_0^t ds \frac{dx}{ds}. \tag{1.18}$$

Since a large deviation principle is proved in general Markov dynamics [2], we know that $V(t)$ satisfies the following asymptotic form in the probability distribution function:

$$\mathrm{Prob}[V(t) = V] \sim e^{-tI(V)} \tag{1.19}$$

with a large deviation function $I(V)$. We show the correspondence between this time-series statics and equilibrium statistical mechanics in Table 1.1. By considering $I(V)$ for negative values of V, we thus find that the large deviation function can give the frequency of the particle climbing up the potential.

Now, let us assume that this system possesses the phenomenological structure for the large deviation principle. This leads to the fact that *the frequency of such a rare event can be obtained phenomenologically without waiting for it.* This property may be used for *a rare event sampling in real experiments*: In many systems, rare events play an important role, such as bio-molecular reactions, nucleation, planetary systems, fully developed turbulence, and plate-tectonic activities. However, the direct observations of these rare events are too demanding. For tackling this obstacle, several techniques for accelerating the observation in numerical simulations have been invented. The examples are transition path sampling [33, 34], transition interface sampling [35, 36], forward flux sampling [37, 38], and the population dynamics method [39, 40]. These are called rare event sampling methods. Until now, as seen in these examples, several rare event sampling methods have been proposed, however, the application of these methods towards *real physical experiments* is still difficult, because most of the rare event sampling methods explained above exploit the property that the experiment is not real but a numerical simulation.[3] Now, we consider the rare event sampling method based on the phenomenological structure for the large deviation principle discussed above. The basic idea of this method comes from *thermodynamics*. Thus, the method should naturally posses a phenomenological feature to apply to real experiments.

This scenario is not restricted within this simple example. For general cases, the same expectation can be made. Since there are a lot of interesting examples of rare events in physics, this rare event sampling method, if it is constructed, will lead to a lot of applications. The second chapter of this thesis is devoted to this attempt.

Before going to the main argument in that chapter, we introduce some results for approaching this problem by using a simple example of the Brownian particle.

1.4.2 Brute Force Approach to the Phenomenological Structure in Time-Series Statistics

By restricting ourselves into the simple example of the Brownian particle, we show how to approach this problem. From the argument in Sect. 1.3, we know that the relation (1.16) is a sufficient condition for the phenomenological structure for the large deviation principle. In time-series statistics, the distribution function of whole configuration, which corresponds to $p_\beta(\Gamma)$ in (1.16), is the path probability density of the particle $p_F[(x(s))_{s=0}^t]$ given as

$$p_F[(x(s))_{s=0}^t] = \frac{1}{C}\exp\left\{-\frac{\gamma}{4T}\int ds\left[\left(\frac{dx(s)}{ds} - \frac{1}{\gamma}F(x(s))\right)^2 + \frac{2T}{\gamma^2}\frac{\partial F(x(s))}{\partial x}\right]\right\}, \tag{1.20}$$

where the multiplication in the integral is interpreted as Stratonovich type [32], and the subscript F represents the tilted periodic potential that appeared in (1.17). By using this path probability density, we define an exponentially biased ensemble $p[(x(s))_{s=0}^t; h]$, which corresponds to $p_\beta(\Gamma; h)$ in (1.16), as

$$p_F[(x(s))_{s=0}^t; h] = \frac{p_F[(x(s))_{s=0}^t]e^{thV(t)}}{\int D[(x(s))_{s=0}^t]p_F[(x(s))_{s=0}^t]e^{thV(t)}}. \tag{1.21}$$

Now, an expectation may be made, it is that the conjugate field of the time-averaged velocity $V(t)$ corresponds directly to the constant force. In other words, we can connect these two path probabilities by

---

[3]For example, in the population dynamics [39, 40], one prepares many of the same initial conditions, then after launching the numerical simulations, the rare-trajectories are copied and the typical-trajectories are killed for each fixed time-intervals. This procedure finally produces trajectories that are quite rare. But, the procedure of copying and killing seems to be difficult to apply in real experiments.

$$p_{F+kh}[(x(s))_{s=0}^{t}] \stackrel{?}{=} p_F[(x(s))_{s=0}^{t}; h] \tag{1.22}$$

with a constant k. This is a direct extension of (1.16) to time-series statistics. However, unfortunately the nature of time-series statistics is more difficult. *The relation* (1.22) *is not true in general.*

In general situations, instead of using a simple constant force kh, we need a special force $\delta w(x)$ that depends on x and h. In order to know the condition determining this special force $\delta w(x)$, we here calculate the ratio between $p_{F+\delta w}[(x(s))_{s=0}^{t}]$ and $p_F[(x(s))_{s=0}^{t}; h]$, which becomes

$$\begin{aligned}
&\frac{p_F[(x(s))_{s=0}^{t}; h]}{p_{F+\delta w}[(x(s))_{s=0}^{t}]} \\
&= \exp\left\{\frac{1}{2T}\int_0^t ds\left[\frac{T}{\gamma}\frac{\partial \delta w(x(s))}{\partial x} + \frac{\delta w(x(s))}{\gamma}F(x(s))\right.\right. \\
&\qquad\left.\left. + \frac{\delta w(x(s))^2}{2\gamma} + (2Th - \delta w(x(s)))\frac{dx(s)}{ds}\right]\right\}.
\end{aligned} \tag{1.23}$$

By setting this ratio to become a constant, we obtain the condition of $\delta w(x)$ for satisfying the corresponding equality to (1.16). By denoting the period of the potential by L, the sufficient condition of $\delta w(x)$ is written as

$$\frac{1}{L}\int_0^L dx \delta w(x) = 2Th \tag{1.24}$$

and

$$\frac{T}{\gamma}\frac{\partial \delta w(x)}{\partial x} + \frac{\delta w(x)}{\gamma}F(x) + \frac{\delta w(x)^2}{2\gamma} = \text{const.} \tag{1.25}$$

We verify that this condition is sufficient as follows: From the first relation, we can eliminate $\int_0^t ds(2Th - \delta w(x(s)))\frac{dx(s)}{ds}$ for large t in (1.23), because this term is not proportional to t when t is large. Then, from the second relation, the remaining terms become a constant. The existence of such a solution $\delta w(x)$ is ensured by Perron-Frobenius theory for liner Matrix [41]. Indeed, by using Cole-Hopf transformation, the equation (1.25) can be linearised. By discretising the space and using a periodic boundary condition, we can map the non-linear problem to a linear Matrix eigenvalue problem. Then, due to the Perron-Frobenius theory, it can be proved that the solution exists and is unique. See the appendix in Ref. [28] for the detail of this argument.

It indicates that we need a special external field determined by (1.24) and (1.25) in order to obtain the phenomenological structure in time-series statistics. This fact is not restricted to this simple Brownian particle system, but is generally true in Markov dynamics. Many similar formulas have been derived [25–31]. For discrete-state-continuous-time Markov dynamics, the corresponding formula was reported in the paper by Jack and Sollich in [26]. For continuous-state-continuous-time Markov dynamics (Langevin dynamics), the corresponding formula was also reported independently by us [27, 28] and Chetrite-Touchette [30, 31]. Mathematically, the formula can be regarded as the generalisation of Doob's h-transform. See Ref. [30] for the details of the explanation.

1.4.3 What Will Be Developed in the Main Chapter of This Thesis

We have seen that we need to solve one dimensional eigenvalue problem even for this simple Brownian dynamics. For more complicated many-body systems, we will see that the situation becomes more demanding as the size of the matrix increases exponentially. In order to use the phenomenological structure for the large deviation principle as a rare-event sampling method, we need a special procedure to construct automatically $\delta w(x)$, instead of relying on the brute-force approach of solving the eigenvalue problem.

In Chap. 2, which is the main chapter in this thesis, we propose such a procedure to obtain $\delta w(x)$. See Fig. 1.4 for the schematic picture of these relations. The key idea for achieving this is *to utilise a measurement instead of solving the eigenvalue problem, to repeat the measurement by dividing the rare-eventness into small pieces*, and *to renormalise this rare-eventness gradually into the system with those repetitive measurements*. In other words, instead of solving the largest eigenvalue problem, we

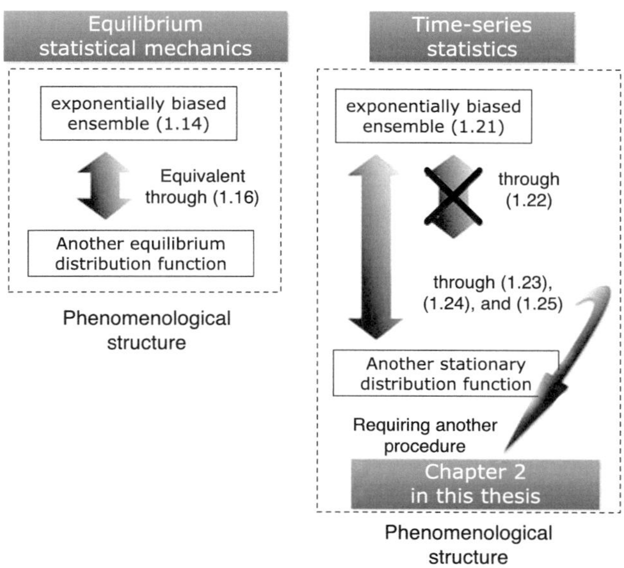

Fig. 1.4 Schematic picture explaining what we will develop in Chap. 2

rely on Monte-Carlo simulation (or real experiment in future), which allows us to obtain this result faster. Then, we make a measurement-and-feedback procedure, where we modify the parameter of the system so as to renormalise rare-eventness gradually into the system after each measurement. We will discuss this procedure in general Markov dynamics in Chap. 2, based on our paper [42].

1.5 Towards Understanding Glass Transition in Terms of Thermodynamic Formalism

Aside from a rare event sampling method, there are many applications of phenomenological structure for the large deviation principle. One of them is to understand glass transition with kinetically constrained models, which is explained below.

Glass transition has an old history, and has been gathering the attention of many researchers. There are several interesting features in glass transition, such as Vogel-Fulcher law for non-Arrenius relaxation of fragile glass, a two-step relaxation of correlation function known as α-relaxation and β-relaxation, the increase of the dynamical correlation length as the system gets close to glass transition point known as dynamical heterogeneity. To explain these specific features of glassy systems, several theories have been studied [43, 44].

For one of the method to approach these problems, a class of models called *kinetically constrained models* has been introduced [21, 43]. This is the coarse grained model of glass formers, which is represented as spin models, where the plus spins and the minus spins are associated with the active and the inactive regions in glass formers. The cooperative properties of glass formers are incorporated as kinetic constraints, for example, the spins surrounded by minus spins can not be flipped up or flipped down. The basic question behind this definition is, does glassy properties come from purely dynamical origin or not? In this model, the stationary distribution is trivial, which is ensured by the detailed balance condition, whereas the dynamics is complicated due to the kinetic constraint explained above. Indeed, this model can explain some features of glasses. For example, the dynamical slowing down, the dynamical heterogeneity [45–51], and an ergodicity breaking transition [52–56]. Furthermore, Garrahan, Jack, Lecomte, Pitard, van Duijvendijk and van Wijland considered recently the large deviation function of a quantity called activity, which represented how much the state of the system changed, and they found that there was a kink in the large deviation function [19–21]. This singularity indicates the existence of *dynamical phase transition*. The singular behaviour appears in infinite system size limit at the origin of that large deviation function (or dynamical free energy in their terminology). Since the slope of the dynamical free energy at the origin corresponds to the expected value of the activity itself, this transition represents that two values of the activity coexist in the infinite system size limit. It is now believed that this coexistence was related to dynamical heterogeneity [21].

As we mentioned above, the dynamical phase transition appears in infinite-size limit. However, one may notice that we can deal with only finite size systems when we study with numerical simulations. Motivated by this fact, after the finding of the dynamical phase transition, finite size scaling method for extracting the property of the dynamical phase transition from finite size systems has been studied, by Bodineau, Toninelli and Lecomte [57, 58]. For constructing it, there is one important problem in thermodynamic formalism: Although we can define an important point for finite size systems as the argument maximising the second-derivative of the dynamical free energy, this point is deviated from the origin in general. When it is deviated from the origin, it is not connected anymore to the static property of the system, because of the lack of physical correspondence in the conjugate field of the activity. For equilibrium magnet substances, there is also a first order phase transition. In this case, the conjugate field of the magnetisation is directly connected to the magnetic field. However, for the case of the dynamical phase transition, there are no direct correspondences between the conjugate field and any physical realistic field.

To this problem, we shall approach with the phenomenological structure in time-series statistics. Like thermodynamics, the physical correspondence to conjugated field should emerge in this structure if it is constructed. In this thesis, we focus on Fredrickson–Andersen (FA) models [59, 60]. We especially study the 1-dimensional version and the mean field version of FA model. (For the detailed definition of the model, refer to the main part of this thesis.) In both cases, we focus on the phenomenological structure in time-series statistics in these models, and investigate some non-trivial properties behind them. Some parts of Chap. 2 are devoted to the 1-dimensional case, and the whole of Chap. 3 is devoted to the mean field case. This work is mainly based on our paper [61].

1.6 Application to van Zon–Cohen Extended Fluctuation Theorem—The Origin in Terms of Rare Trajectories

Here, as another application of the phenomenological structure for the large deviation principle, we consider an extended fluctuation theorem reported by van Zon and Cohen [62, 63].

The fluctuation theorem is a symmetry property of the fluctuation of the entropy production [25], from which several confirmed results, e.g. second law of thermodynamics and linear response formula, were derived very beautifully (See for example, Ref. [64]). The first experimental verification of this theorem was done by Wang et al. [65] in 2001. They used a Brownian particle dragged by an optical tweezer, and measured the work fluctuation in order to verify the fluctuation theorem for the work. The detailed analysis of their set up was done by van Zon and Cohen in [66]. Thanks to these works, the fluctuation theorem was confirmed in terms of both of the experimental and the theoretical points of view. However, by extending their method, van Zon and Cohen submitted another paper later that showed that the fluctuation theorem with respect to heat was different [62, 63]. Heat fluctuation did not satisfy the fluctuation theorem, but it satisfied a different symmetry. Until now, several features about this new symmetry have been studied [67–74]. However, the direct understanding of it with respect to the rare trajectories of the particle has not been explored yet. We tackle this problem by using the phenomenological structure for the large deviation principle. This work is based on our paper [75]. Chapter 4 is devoted to this attempt.

1.7 Construction of Thesis

This thesis is composed of four chapters. *Introduction*, Chap. 1, was presented above. Chapter 2, which is the main chapter in this thesis, is *Iterative measurement-feedback procedure for large deviation statistics*. In this chapter, we propose a computational method for large deviation statistics, where the procedure is constituted of only measurements and feedbacks. Since this procedure is a direct extension of thermodynamics to time-series statistics, we believe that this method will become the first rare-event sampling method used in real experiments in future. This chapter corresponds to our paper [42]. Chapter 3 is *Common scaling function in dynamical and quantum phase transitions*. In this chapter, inspired by the finite-size scaling method of first order phase transition for equilibrium spin models [76, 77], we propose a finite-size scaling method for the dynamical phase transitions. Then, it turns out that the method can also be applied to quantum phase transitions. We confirm this method in mean-field models, however, we believe that this method is general and becomes a wide-ranging prescription for analysing dynamical and quantum phase transitions. This chapter corresponds to our paper [61]. In the next chapter, Chap. 4, we discuss the application of the phenomenological structure to van-Zon extended fluctuation theorem. This chapter corresponds to our paper [75]. Finally, Chap. 5 is *Conclusions and future perspectives*, where we show some future open problems.

References

1. C.H. Callen, *Thermodynamics and an Introduction to Thermostatistics*, 2nd edn. (Wiley, New York, 1985)
2. A. Dembo, O. Zeitouni, *Large Deviations Techniques and Applications* (Springer, New York, 1998)
3. H. Touchette, Phys. Rep. **478**, 1 (2009)
4. D. Ruelle, *Thermodynamic Formalism* (Addison-Wesley, Reading, 1978)
5. D.J. Evans, E.G.D. Cohen, G.P. Morriss, Phys. Rev. Lett. **71**, 2401 (1993)
6. G. Gallavotti, E.G.D. Cohen, Phys. Rev. Lett. **74**, 2694 (1995)
7. J. Kurchan, J. Phys. A **31**, 3719 (1998)
8. C. Maes, J. Stat. Phys. **95**, 367 (1999)
9. G.E. Crooks, Phys. Rev. E **60**, 2721 (1999)
10. J.L. Lebowitz, H. Spohn, J. Stat. Phys. **95**, 333 (1999)

11. B. Derrida, J.L. Lebowitz, Phys. Rev. Lett. **80**, 209 (1998)
12. B. Derrida, Phys. Rep. **301**, 65 (1998)
13. T. Bodineau, B. Derrida, Phys. Rev. Lett. **92**, 180601 (2004)
14. T. Bodineau, B. Derrida, J. Stat. Phys. **123**, 277 (2006)
15. T. Bodineau, B. Derrida, C. R. Phys. **8**, 540 (2007)
16. B. Derrida, J. Stat. Mech. (2007) P07023
17. L. Bertini, A. De Sole, D. Gabrielli, G. Jona-Lasinio, C. Landim, Phys. Rev. Lett. **94**, 030601 (2005)
18. L. Bertini, A. De Sole, D. Gabrielli, G. Jona-Lasinio, C. Landim, J. Stat. Phys. **123**, 237 (2006)
19. J.P. Garrahan, R.L. Jack, V. Lecomte, E. Pitard, K. van Duijvendijk, F. van Wijland, Phys. Rev. Lett. **98**, 195702 (2007)
20. J.P. Garrahan, R.L. Jack, V. Lecomte, E. Pitard, K. van Duijvendijk, F. van Wijland, J. Phys. A **42**, 075007 (2009)
21. J.P. Garrahan, P. Sollich, C. Toninelli, (2010). arXiv:1009.6113, which is a chapter in Ref. L. Berthier, G. Biroli, J.-P. Bouchaud, L. Cipelletti, W. van Saarloos, (eds.), *Dynamical Heterogeneities in Glasses, Colloids, and Granular Media* (Oxford University Press, Oxford, 2011)
22. C. Maes, K. Netočný, B. Wynants, Phys. Rev. Lett. **107**, 010601 (2011)
23. A. Lazarescu, K. Mallick, J. Phys. A: Math. Theor. **44**, 315001 (2011)
24. M. Gorissen, A. Lazarescu, K. Mallick, C. Vanderzande, Phys. Rev. Lett. **109**, 170601 (2012)
25. R.M.L. Evans, Phys. Rev. Lett. **92**, 150601 (2004)
26. R.L. Jack, P. Sollich, Prog. Theor. Phys. Suppl. **184**, 304 (2010)
27. T. Nemoto, S. Sasa, Phys. Rev. E **83**, 030105(R) (2011)
28. T. Nemoto, S. Sasa, Phys. Rev. E **84**, 061113 (2011)
29. S. Sasa, Phys. Scr. **86**, 058514-1-3 (2012)
30. R. Chetrite, H. Touchette, Phys. Rev. Lett. **111**, 120601 (2013)
31. R. Chetrite, H. Touchette, (2014). arXiv:1405.5157
32. C.W. Gardiner, *Handbook of Stochastic Methods for Physics, Chemistry, and the Natural Sciences* (Springer, Berlin, 1983)
33. D. Frenkel, B. Smit, *Understanding Molecular Simulation* (Academic Press, San Diego, 2001)
34. P.G. Bolhuis, D. Chandler, C. Dellago, P.L. Geissler, Annu. Rev. Phys. Chem. **53**, 291 (2002)
35. T.S. van Erp, D. Moroni, P.G. Bolhuis, J. Chem. Phys. **118**, 7762 (2003)
36. T.S. van Erp, P.G. Bolhuis, J. Comput. Phys. **205**, 157 (2005)
37. R.J. Allen, P.B. Warren, P.R. ten Wolde, Phys. Rev. Lett. **94**, 018104 (2005)
38. R.J. Allen, D. Frenkel, P.R. ten Wolde, J. Chem. Phys. **124**, 024102 (2006)
39. C. Giardina, J. Kurchan, L. Peliti, Phys. Rev. Lett. **96**, 120603 (2006)
40. C. Giardina, J. Kurchan, V. Lecomte, J. Tailleur, J. Stat. Phys. **145**, 787 (2011)
41. E. Seneta, *Non-Negative Matrices and Markov Chains*, 2nd edn. (Springer, New York, 2006)
42. T. Nemoto, S. Sasa, Phys. Rev. Lett. **112**, 090602 (2014)
43. F. Ritort, P. Sollich, Adv. Phys. **52**, 219 (2003)
44. P.G. Debenedetti, F.H. Stillinger, Nature (London) **410**, 259 (2001)
45. J.P. Garrahan, D. Chandler, Phys. Rev. Lett. **89**, 035704 (2002)
46. C. Toninelli, G. Biroli, D.S. Fisher, Phys. Rev. Lett. **92**, 185504 (2004)
47. S. Whitelam, L. Berthier, J.P. Garrahan, Phys. Rev. Lett. **92**, 185705 (2004)
48. A.C. Pan, J.P. Garrahan, D. Chandler, Phys. Rev. E **72**, 041106 (2005)
49. C. Toninelli, M. Wyart, L. Berthier, G. Biroli, J.P. Bouchaud, Phys. Rev. E **71**, 041505 (2005)
50. L. Berthier, J.P. Garrahan, J. Phys. Chem. B **109**, 3578 (2005)
51. R.L. Jack, P. Mayer, P. Sollich, J. Stat. Mech. (2006) P03006
52. S.N. Dorogovtsev, A.V. Goltsev, J.F.F. Mendes, Phys. Rev. Lett. **96**, 040601 (2006)
53. J.M. Schwarz, A.J. Liu, L.Q. Chayes, Europhys. Lett. **73**, 560 (2006)
54. J. Reiter, F. Mauch, J. Jäckle, 1992. Physica A **184**, 458 (1992)
55. C. Toninelli, G. Biroli, D.S. Fisher, Phys. Rev. Lett. **98**, 129602 (2007)
56. C. Toninelli, G. Biroli, Eur. Phys. J. B **130**, (2008)
57. T. Bodineau, C. Toninelli, Commun. Math. Phys. **311**, 357 (2012)

58. T. Bodineau, V. Lecomte, C. Toninelli, J. Stat. Phys. **147**, 1 (2012)
59. G.H. Fredrickson, H.C. Andersen, Phys. Rev. Lett. **53**, 1244 (1984)
60. G.H. Fredrickson, H.C. Andersen, J. Chem. Phys. **83**, 5822 (1985)
61. T. Nemoto, V. Lecomte, S. Sasa, F. van Wijland, J. Stat. Mech. (2014) P10001
62. R. van Zon, E.G.D. Cohen, Phys. Rev. Lett. **91**, 110601 (2003)
63. R. van Zon, E.G.D. Cohen, Phys. Rev. E **69**, 056121 (2004)
64. K. Hayashi, S. Sasa, Physica A **370**, 407 (2006)
65. G.M. Wang, E.M. Sevick, E. Mittag, D.J. Searles, D.J. Evans, Phys. Rev. Lett. **89**, 050601 (2002)
66. R. van Zon, E.G.D. Cohen, Phys. Rev. E **67**, 046102 (2003)
67. N. Garnier, S. Ciliberto, Phys. Rev. E **71**, 060101(R) (2005)
68. F. Bonetto, G. Gallavotti, A. Giuliani, F. Zamponi, J. Stat. Phys. **123**, 39 (2006)
69. M. Baiesi, T. Jacobs, C. Maes, N.S. Skantzos, Phys. Rev. E **74**, 021111 (2006)
70. P. Visco, J. Stat. Mech. (2006) P06006
71. R.J. Harris, A. Rákos, G.M. Schütz, Europhys. Lett. **75**, 227 (2006)
72. R.J. Harris, A. Rákosand, J. Stat. Mech. (2008) P05005
73. A. Puglisi, L. Rondoni, A. Vulpiani, J. Stat. Mech. (2006) P08010
74. J.D. Noh, J.-M. Park, Phys. Rev. Lett. **108**, 240603 (2012)
75. T. Nemoto, Phys. Rev. E **85**, 061124 (2012)
76. C. Borgs, R. Kotecký, J. Stat. Phys. **61**, 79 (1990)
77. C. Borgs, R. Kotecký, Phys. Rev. Lett. **68**, 1734 (1992)

Chapter 2
Iterative Measurement-Feedback Procedure for Large Deviation Statistics

2.1 Introduction

In the last two decades, large deviation functions in time-series statistics have gathered attention in the field of nonequilibrium physics. The beginning is the discovery of the fluctuation theorem, which is the symmetry property of the large deviation function of the time-averaged entropy production rate [1–6]. After that, several results for the large deviation functions have followed, such as an additivity principle for driven diffusive systems [7–12], generalised Onsager–Machlup approach [13, 14], dynamical phase transitions of kinetically constrained models [15–17], a Lyapunov function for non-equilibrium steady states without relying on entropy production [18], exact results for the current statistics of lattice gas models [17, 19].

Also, with these developments, there were some studies that focused on the thermodynamic structure in time-series statistics [20–26]. Especially in these analysis, a technique to map a biased ensemble to another steady state ensemble has been utilised. This mapping is well defined in mathematical sense, however, in order to construct this mapping, we need pre-information for generating a special external force added. That requires to solve eigenvalue problems of a matrix in a large dimension: the degrees of freedom of the system, which is demanding in computational point of view. For overcoming this difficulty, we proposed a variational principle constituted of observable quantities to determine that external force [24–26]. Since the variational parameter of this variational principle was an external field of the system, it offers a simple method for determining that external force without solving any mathematical largest eigenvalue problems. However, there is still a problem for large size systems, because the domain of the variational functional increases exponentially as the system size becomes larger in general.

Here in this chapter, we propose a new computational method for large deviation statistics. This exploits a property in time-series statistics, which is *an additive property of the rareeventness with a special measurement and feedback.*: By iterating a procedure constituted of measurements and feedbacks, we gradually renormalise

T. Nemoto, *Phenomenological Structure for the Large Deviation Principle in Time-Series Statistics*, Springer Theses,
DOI 10.1007/978-981-287-811-3_2

the system, and let the obtained modified systems attain the rare-event property. We stress that this method is constituted of measurements and feedbacks. Thus, it can be implemented in real experiment in principle. Furthermore, by combining it with an idea of *effective description of exponential family*, we show a good numerical performance of the method. Indeed, as a demonstration, we apply this method to many-body systems, and obtain some non-trivial features of the rare fluctuations in those systems.

The construction of this chapter is as follows. In Sect. 2.2, we show some preliminaries. In Sect. 2.2.1–2.2.4, we give the definition of the model and some basics of a large deviation principle. Then, in Sect. 2.2.5, we show and prove a mapping method from a biased ensemble to the steady state dynamics, which is the key formula to construct the phenomenological structure in time-series statistics. In Sect. 2.3, we explain our computational method. From Sect. 2.3.1–2.3.3, we explain the idea behind the method, and in Sect. 2.3.4, we show explicitly the procedure of the method. Section 2.4 is devoted to the application of the method to non-equilibrium many-body lattice gas models. In particular, in Sect. 2.4.1, we introduce an effective description of the exponential family, and in the following subsections, with the effective description, we analyse these many-body systems. Finally, in Sect. 2.5, we make a conclusion of this chapter. This chapter is based on the paper published in [27].

2.2 Preliminary

2.2.1 Model

The state space Ω is a finite set. On Ω, we consider continuous time Markov processes. For $\boldsymbol{n}, \boldsymbol{n}' \in \Omega$, we define a transition rate $w(\boldsymbol{n} \to \boldsymbol{n}')$ as an irreducible matrix that satisfies $w(\boldsymbol{n} \to \boldsymbol{n}) = 0$ and $w(\boldsymbol{n} \to \boldsymbol{n}') \neq 0$ if $w(\boldsymbol{n}' \to \boldsymbol{n}) \neq 0$. The escape rate is defined as

$$\lambda(\boldsymbol{n}) \equiv \sum_{\boldsymbol{n}' \in \Omega} w(\boldsymbol{n} \to \boldsymbol{n}'). \tag{2.1}$$

We start with an initial distribution function $P_0(\boldsymbol{n})$. Then, the distribution function of $\boldsymbol{n}$ at time t, $P(\boldsymbol{n}) \equiv \langle \delta_{\boldsymbol{n}(t),\boldsymbol{n}} \rangle$, is determined from the following Master equation [28]:

$$\frac{\partial}{\partial t} P(\boldsymbol{n}, t) = \sum_{n} P(\boldsymbol{n}) w(\boldsymbol{n} \to \boldsymbol{n}') - \delta_{\boldsymbol{n},\boldsymbol{n}'} \lambda(\boldsymbol{n}). \tag{2.2}$$

We denote the history of states during a time interval t by ω, which is specified by the total number of transitions n, a collection of transition times $(t_i)_{i=1}^{n}$, and a sequence of states $(\boldsymbol{n}_i)_{i=0}^{n}$, where $\boldsymbol{n}_i = \boldsymbol{n}(t)$ for $t_i \le t \le t_{i+1}$ with $t_0 \equiv 0$, $t_{n+1} \equiv t$. See Fig. 2.1 for the schematic diagram explaining this definition. We denote the path probability density with an initial condition $\boldsymbol{n}_0$ by $P(\omega|\boldsymbol{n}_0)$. It becomes

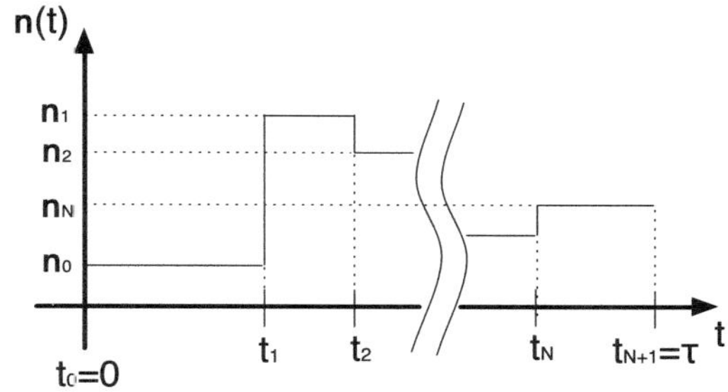

Fig. 2.1 The schematic diagram to explain the definition of the history of states, ω

$$P(\omega|\boldsymbol{n}_0) = e^{-\lambda(\boldsymbol{n}_0)t_1} \prod_{i=1}^{N} \left[w(\boldsymbol{n}_{i-1} \to \boldsymbol{n}_i) e^{-\lambda(\boldsymbol{n}_i)(t_{i+1}-t_i)} \right] \tag{2.3}$$

or equivalently,

$$P(\omega|\boldsymbol{n}_0) = e^{-\int_0^t d\tilde{t}\lambda(\boldsymbol{n}(\tilde{t}))} \prod_{i=1}^{N} \left[w(\boldsymbol{n}_{i-1} \to \boldsymbol{n}_i) \right] \tag{2.4}$$

For the sake of completeness, we show a simple derivation of the path probability in Appendix A.1.

2.2.2 Cumulant Generating Function

For each transition $\boldsymbol{n}_i \to \boldsymbol{n}_{i+1}$, we define a quantity $\alpha(\boldsymbol{n}_i \to \boldsymbol{n}_{i+1})$. Then, we consider the corresponding time-averaged value $A(\omega)$, which is defined as

$$A(\omega) = \frac{1}{\tau} \sum_{i=0}^{n-1} \alpha(\boldsymbol{n}_i \to \boldsymbol{n}_{i+1}). \tag{2.5}$$

Since the system is Markovian, $A(\omega)$ has a large deviation principle in the limit of $\tau \to \infty$. That is, with a probability density for $A(\omega)$, $p(A)$, shows the following asymptotic form:

$$p(A) \sim e^{-\tau I(A)}, \tag{2.6}$$

where $I(A)$ is a large deviation function. We note that the expected value or typical value of $A(\omega)$, $\langle A(\omega) \rangle$ is determined by a variational principle

$$\langle A(\omega) \rangle = \underset{A}{\text{Argmin}} \left[I(A) \right]. \tag{2.7}$$

Here, we introduce a scaled cumulant generating function defined by

$$G(h) \equiv \lim_{\tau\to\infty} \frac{1}{\tau} \log \left\langle e^{h\tau A(\omega)} \right\rangle, \tag{2.8}$$

where h is called a biasing parameter. It resembles the definition of the Helmholtz free energy in equilibrium statistical mechanics, so that this function is called a dynamical free energy. Similar to equilibrium thermodynamics, the large deviation function $I(A)$ and the cumulant generating function is connected through Legendre transformation:

$$I(A) = \max_h [hA - G(h)], \tag{2.9}$$

or

$$G(h) = \max_A [hA - I(A)]. \tag{2.10}$$

We should mention that we are now considering the case that $I(A)$ is a concave function. Otherwise, this relationship can be broken. See Ref. [29] for the explanation of the example.

2.2.3 Biased Ensemble

In the analysis of a large deviation principle, a class of modified pass probability measure called *an exponential family* or *a biased ensemble*, $P(\omega; h)$, is often studied. With a parameter h representing how much the system is biased, $P(\omega; h)$ is defined as

$$P(\omega; h) = \frac{1}{Z(\tau, h)} P(\omega) e^{h\tau A(\omega)}, \tag{2.11}$$

where $Z(\tau, h)$ is the normalisation constant defined by $Z(\tau, h) = \left\langle e^{h\tau A(\omega)} \right\rangle$. In this biased ensemble, rare trajectories characterised by the large deviation principle of $A(\omega)$ has a large (or small) probability compared with the one in the original ensemble. Indeed, by denoting the probability density of $A(\omega)$ in the biased ensemble by $p(A; h)$, we obtain

$$p(A; h) = \frac{1}{Z(\tau, h)} p(A) e^{h\tau A} \sim \frac{1}{Z(\tau, h)} e^{-\tau(I(A) - hA)}. \tag{2.12}$$

for large τ, where we used the large deviation principle of $A(\omega)$. We first notice that the large deviation function of $A(\omega)$ in the biased ensemble is given as $I(A) - hA$. Then, the following result directly yields: the expected value (or typical value) of A in the biased ensemble is determined by the variational principle

$$\frac{\left\langle A(\omega)e^{h\tau A(\omega)}\right\rangle}{Z(\tau,h)} = \underset{A}{\text{Argmin}}\left[I(A) - hA\right]. \tag{2.13}$$

The typical value is thus deviated from $\langle A(\omega)\rangle$. Furthermore it is connected to the cumulant generating function through

$$\frac{\partial G(h)}{\partial h} = \lim_{\tau\to\infty}\frac{\left\langle A(\omega)e^{h\tau A}\right\rangle}{Z(\tau,h)}. \tag{2.14}$$

Therefore, due to the equivalence between $G(h)$ and $I(A)$, we find a relation connecting the expected value in the biased ensemble with the large deviation function:

$$I(A) = \max_h\left[hA - \int_0^h d\tilde{h}\lim_{\tau\to\infty}\frac{\left\langle A(\omega)e^{\tilde{h}\tau A(\omega)}\right\rangle}{Z(\tau,\tilde{h})}\right]. \tag{2.15}$$

2.2.4 *Revisit of the Phenomenological Structure for the Large Deviation Principle in Equilibrium Statistical Mechanics*

Here, let us revisit the phenomenological structure for the large deviation principle in equilibrium statistical mechanics with a viewpoint of biased ensemble. The biased ensemble (2.11) reminds us of the canonical distribution function (given in (1.6)). We here show that this similarity leads to the key of the phenomenological structure for the large deviation principle in equilibrium statistical physics.

In the system considered in the introduction 1.2, we define the biased ensemble $p(\Gamma; h)$ by

$$p(\Gamma; h) = \frac{1}{Z(h)}p(\Gamma)e^{hH(\Gamma)}. \tag{2.16}$$

Then, we obtain the distribution function of U in the biased ensemble as

$$p(U; h) = \frac{1}{Z(h)}\Omega(U)e^{-(\beta-h)U} \sim \frac{1}{\tilde{Z}(h)}e^{N[s(u)-(\beta-h)u]}. \tag{2.17}$$

Because no β dependence appears here in the first term $s(u)$ of this exponential function, this equation indicates that the biased ensemble is the equilibrium distribution function of the system with temperature $\beta - h$. Thus the biasing parameter h is renormalised as the temperature of the system in another equilibrium system. The same structure, namely the correspondence between biasing parameters and equilibrium intensive parameters, is true not only for the energy density discussed here, but also for the density of general thermodynamic extensive quantities.

This is the key for the phenomenological structure in equilibrium statistical mechanics. Indeed, from this relation, we can connect the large deviation function

$I(U)$ with the expected value of U in another equilibrium system, which has a new temperature $\beta' = \beta - h$. Indeed, by denoting this expected value by $\langle U \rangle_{\beta'=\beta-\tilde{h}}$ and using the corresponding equation to (2.15) in equilibrium statistical mechanics, we have

$$I(U) = \max_h \left[hU - \int_0^h d\tilde{h} \, \langle U \rangle_{\beta'=\beta-\tilde{h}} \right]. \tag{2.18}$$

In short, the key to the phenomenological structure for the large deviation principle is the physical correspondence of the biased ensemble. Because the biased ensemble corresponds to another equilibrium system, we can construct a large deviation function from expected values.

2.2.5 Steady Dynamics Corresponding to Biased Ensemble

The phenomenological structure for the large deviation principle in equilibrium statistical mechanics is clarified in the previous subsection, where we found what we need to construct the same structure in time-series statistics: It is the construction of a physical system corresponding to the biased ensemble (2.11) in time-series statistics. Now, in order to achieve this construction, we follow the following strategy:

1. We define a new transition rate

$$w^h(\boldsymbol{n} \to \boldsymbol{n}') \equiv w(\boldsymbol{n} \to \boldsymbol{n}') f^h(\boldsymbol{n} \to \boldsymbol{n}') \tag{2.19}$$

   with an unknown function $f^h(\boldsymbol{n} \to \boldsymbol{n}')$ that depends on h. Then, the new path probability $P^h(\omega|\boldsymbol{n}_0)$ is given as

$$P^h(\omega|\boldsymbol{n}_0) = e^{-\int_0^t d\tilde{t} \lambda^h(\boldsymbol{n}(\tilde{t}))} \prod_{i=1}^{N} \left[w(\boldsymbol{n}_{i-1} \to \boldsymbol{n}_i) \right] \prod_{i=1}^{N} \left[f^h(\boldsymbol{n}_{i-1} \to \boldsymbol{n}_i) \right], \tag{2.20}$$

   where we defined $\lambda^h(\boldsymbol{n}) \equiv \sum_{\boldsymbol{n}'} w^h(\boldsymbol{n} \to \boldsymbol{n}')$.
2. We consider the ratio between this new path probability density and the biased path probability density $P(\omega; h|\boldsymbol{n}_0)$,

$$\frac{P^h(\omega|\boldsymbol{n}_0)}{P(\omega; h|\boldsymbol{n}_0)} = e^{-\int_0^t d\tilde{t} [\lambda^h(\boldsymbol{n}(\tilde{t})) - \lambda(\boldsymbol{n}(\tilde{t}))]} \prod_{i=1}^{N} \left[f^h(\boldsymbol{n}_{i-1} \to \boldsymbol{n}_i) e^{-h\alpha(\boldsymbol{n}_{i-1} \to \boldsymbol{n}_i)} \right] Z(t, h). \tag{2.21}$$

   Then, we determine $f^h(\boldsymbol{n} \to \boldsymbol{n}')$ so as to make the right-hand side of (2.21) a constant. This construction leads to the conclusion that the new system characterised by $w^h(\boldsymbol{n} \to \boldsymbol{n}')$ has the same path probability density as the one in this modified dynamics. In the following part, we show how to determine $f^h(\boldsymbol{n} \to \boldsymbol{n}')$.

3. First, we look at the product part ($\prod[\cdots]$). For obtaining a benefit from this product structure, we set

$$f^h(\boldsymbol{n} \to \boldsymbol{n}') = e^{h\alpha(\boldsymbol{n}\to\boldsymbol{n}')} \frac{\phi(\boldsymbol{n}')}{\phi(\boldsymbol{n})} \tag{2.22}$$

with an unknown function $\phi(\boldsymbol{n})$. In the product from $i = 1$ to $i = N$, a ratio in this right-hand side will be canceled each other in total. Indeed, the total product becomes

$$\prod_{i=1}^{N} \left[f^h(\boldsymbol{n}_{i-1} \to \boldsymbol{n}_i) e^{-h\alpha(\boldsymbol{n}_{i-1}\to\boldsymbol{n}_i)} \right] = \frac{\phi(\boldsymbol{n}_N)}{\phi(\boldsymbol{n}_0)}. \tag{2.23}$$

Since this right-hand side is small compared with the other parts in large τ limit, we can neglect it in (2.21).
Second, we determine the unknown function $\phi(\boldsymbol{n})$ so as to make the first exponential part in (2.21) a constant. For this purpose, we set a condition to $\phi(\boldsymbol{n})$ as

$$\lambda^h(\boldsymbol{n}) - \lambda(\boldsymbol{n}) = \text{const.} \equiv K, \tag{2.24}$$

which indeed ensures that that first exponential part a constant e^{-tK}. This equation can be rewritten as an eigenvalue problem of an irreducible matrix

$$L^h_{\boldsymbol{n}',\boldsymbol{n}} = w(\boldsymbol{n} \to \boldsymbol{n}') e^{h\alpha(\boldsymbol{n}\to\boldsymbol{n}')} - \delta_{\boldsymbol{n},\boldsymbol{n}'} \lambda(\boldsymbol{n}). \tag{2.25}$$

Indeed, we rewrite (2.24) as

$$\sum_{\boldsymbol{n}'} w(\boldsymbol{n} \to \boldsymbol{n}') e^{h\alpha(\boldsymbol{n}\to\boldsymbol{n}')} \phi(\boldsymbol{n}') - \lambda(\boldsymbol{n})\phi(\boldsymbol{n}) = K\phi(\boldsymbol{n}), \tag{2.26}$$

which is equivalent to

$$\sum_{\boldsymbol{n}'} \phi(\boldsymbol{n}') L^h_{\boldsymbol{n}',\boldsymbol{n}} = K\phi(\boldsymbol{n}). \tag{2.27}$$

Here, we remember that we need to impose a condition that the eigenfunction $\phi(\boldsymbol{n})$ is a positive vector, because the transition matrix $w^h(\boldsymbol{n} \to \boldsymbol{n}')$ can not take a negative value. We thus find that the K is the largest eigenvalue of $L^h_{\boldsymbol{n}',\boldsymbol{n}}$ due to Perron-Frobenius theory [30]. Also, we note that the largest eigenvalue K and the corresponding left-eigenvector $\phi(\boldsymbol{n})$ surely exist and are unique, due to this theory.

Therefore, by choosing $\phi(\boldsymbol{n})$ as the left-eigenvector of the largest eigenvalue of $L^h_{\boldsymbol{n}',\boldsymbol{n}}$, we will reach the desired result.

Many formulas similar to this have been derived [20–26]. Exactly the same form as this result was reported in the paper by Jack and Sollich in [23]. For

Langevin systems, the corresponding formulas were also derived independently by us [24, 25] and Chetrite, Touchette [20, 21]. Mathematically, the formula is regarded as the generalisation of Doob's h-transform. See Ref. [21] for the details of the explanation.

We mention that the corresponding system is characterised by a variational principle [24–26]. With a variational functional $\tilde{V}(\boldsymbol{n})$, we introduce a variational transition rate $\tilde{w}_h^{\tilde{V}}(\boldsymbol{n} \to \boldsymbol{n}')$ as

$$\tilde{w}_h^{\tilde{V}}(\boldsymbol{n} \to \boldsymbol{n}') = w(\boldsymbol{n} \to \boldsymbol{n}')e^{h\alpha(\boldsymbol{n}\to\boldsymbol{n}')-(1/2)\tilde{V}(\boldsymbol{n}')+(1/2)\tilde{V}(\boldsymbol{n})}. \tag{2.28}$$

Also, we denote the expected value in the stationary state generated by $\langle\ \rangle_h^{\tilde{V}}$, and the escape rate in the system $\tilde{w}_h^{\tilde{V}}(\boldsymbol{n} \to \boldsymbol{n}')$ by $\tilde{\lambda}_h^{\tilde{V}}$. Then, that variational principle is written as

$$w^h(\boldsymbol{n} \to \boldsymbol{n}') = \operatorname*{Argmax}_{\tilde{V}} \left\langle \tilde{\lambda}^{\tilde{V}} - \lambda \right\rangle_h^{\tilde{V}} \tag{2.29}$$

and the maximum value gives the cumulant generating function itself

$$G(h) = \max_{\tilde{V}} \left\langle \tilde{\lambda}^{\tilde{V}} - \lambda \right\rangle_h^{\tilde{V}}. \tag{2.30}$$

This variational principle was studied by us with a motivation to construct the corresponding steady state from *observable quantities of the system* [25]. Indeed, if we apply it to Langevin equation, the escape rate is replaced by an entropy production rate and the variational potential corresponds to the real potential added to the Brownian particle. The mathematical origin of the variational principle is different from the thermodynamic one. Rather, that variational principle is related to Donsker–Varadhan formula for empirical measure [31]. See Appendix A.2 for the derivation of the variational principle from Donsker–Varadhan formula. Furthermore, when the system satisfies detailed balance condition, the variational principle can be connected to the one in quantum mechanics. This is explained in the Chap. 3, where we apply the variational principle to a kinetically constrained model for deriving a scaling function around the phase transition.

Here, we also mention that there is the case that the derivation above is not correct, where the boundary term $\phi(\boldsymbol{n}_N)/\phi(\boldsymbol{n}_0)$ in (2.23) can not be neglected. This problem is one of the origin of the extended fluctuation theorem of heat dissipation, reported by van Zon and Cohen [32, 33]. In Chap. 4, we discussed this connection explicitly.

As a corollary of this formulation, we obtain an equivalence between the largest eigenvalue K and the cumulant generating function $G(h)$:

$$K = G(h). \tag{2.31}$$

Indeed, by using (2.23) and (2.24) in (2.21), we have

$$P^h(\omega|\boldsymbol{n}_0) = P(\omega; h|\boldsymbol{n}_0)e^{-Kt}\frac{\phi(\boldsymbol{n}_N)}{\phi(\boldsymbol{n}_0)}Z(t,h) \tag{2.32}$$

By taking the sum with respect to ω and considering only the dominant term in large t limit, we obtain

$$K = \lim_{t\to\infty}\frac{1}{t}\log Z(t,h), \tag{2.33}$$

which is (2.31). The cumulant generating function is obtained from the largest eigenvalue problem of $L^h_{\boldsymbol{n},\boldsymbol{n}'}$, and then the result is connected to the large deviate function through Legendre transformation (2.9). Because it is easier to deal with the largest eigenvalue problem than the large deviation principle itself, this structure has been used in many situation for mathematically rigorous analysis in large deviation theory [34]. Furthermore, the relation has been used for analysing non equilibrium systems. The example includes the fluctuation theorem by Lebowitz and Spohn [5], where they found a symmetry property in cumulant generating function of entropy production rate due to the fluctuation theorem.

2.3 Main Result

In the previous subsection, we finally understood how we could create the corresponding system to the biased ensemble, where we needed to solve the largest eigenvalue problem of $L^h_{\boldsymbol{n}',\boldsymbol{n}}$. For many body systems, however, it is demanding and almost impossible to solve the corresponding largest eigenvalue problem because the number of the degrees of the freedom in these systems increases exponentially with the system size. Here in order to overcome this difficulty, we propose a method to obtain the corresponding system with measurements and feedbacks.

2.3.1 *Measurement Formula of $\phi(\boldsymbol{n})$ in Monte-Carlo Simulations*

Because the direct diagonalisation of the largest eigenvalue of $L^h_{\boldsymbol{n}',\boldsymbol{n}}$ is hopeless, we rely on Monte-Carlo simulations instead. First, we show a method to obtain $\phi(\boldsymbol{n})$ from Monte-Carlo simulations.

First, we define $\tilde{\psi}(\boldsymbol{n}, t|\boldsymbol{n}_0)$ obtained from the following initial condition and evolution equation:

$$\tilde{\psi}(\boldsymbol{n}, 0|\boldsymbol{n}_0) = \delta_{\boldsymbol{n},\boldsymbol{n}_0} \tag{2.34}$$

$$\frac{\partial\tilde{\psi}(\boldsymbol{n}, t|\boldsymbol{n}_0)}{\partial t} = \sum_{\boldsymbol{n}'}\tilde{\psi}(\boldsymbol{n}', 0|\boldsymbol{n}_0)L^h_{\boldsymbol{n},\boldsymbol{n}'}. \tag{2.35}$$

Then, $\tilde{\psi}(\boldsymbol{n}, t|\boldsymbol{n}_0)$ equals to $\left\langle \delta_{\boldsymbol{n}(t),\boldsymbol{n}} e^{htA(\omega)} \right\rangle_{\boldsymbol{n}_0}$, where $\langle\ \rangle_{\boldsymbol{n}_0}$ is the expected value with respect to the Monte-Carlo simulation with an initial condition $\boldsymbol{n}(0) = \boldsymbol{n}_0$:

$$\tilde{\psi}(\boldsymbol{n}, t|\boldsymbol{n}_0) = \left\langle \delta_{\boldsymbol{n}(t),\boldsymbol{n}} e^{htA(\omega)} \right\rangle_{\boldsymbol{n}_0}. \tag{2.36}$$

The way to prove this is to show that $\left\langle \delta_{\boldsymbol{n}(t),\boldsymbol{n}} e^{htA(\omega)} \right\rangle_{\boldsymbol{n}_0}$ also satisfies (2.34) and (2.35). This is done in Appendix A.3. Next, since $L^h_{\boldsymbol{n}',\boldsymbol{n}}$ is irreducible, the large time behaviour of $\tilde{\psi}(\boldsymbol{n}, t|\boldsymbol{n}_0)$ is

$$\tilde{\psi}(\boldsymbol{n}, t|\boldsymbol{n}_0) \sim \phi(\boldsymbol{n}_0)\psi(\boldsymbol{n})e^{tK}. \tag{2.37}$$

with a definition of $\psi(\boldsymbol{n})$ as the right-largest eigenvector of $L^h_{\boldsymbol{n}',\boldsymbol{n}}$. Thus, by combining (2.36) with (2.37), we arrive at

$$\phi(\boldsymbol{n}) \propto \left\langle e^{htA(\omega)} \right\rangle_{\boldsymbol{n}} \tag{2.38}$$

for large t. This is a basic result that directly came from the largest eigenvalue analysis. Thanks to (2.38), we can reach $\phi(\boldsymbol{n})$ just by using a Monte Carlo simulation in principle, however, we will face a difficulty with this formula soon, which is explained next.

2.3.2 Rare Events Required for Measurement of $\left\langle e^{hTA(\omega)} \right\rangle_{\boldsymbol{n}}$

Even though we obtain (2.38), it is easy to show that the dominant contribution of the ensemble to obtain $\left\langle e^{htA(\omega)} \right\rangle_{\boldsymbol{n}}$ is rare, which is characterised by a large deviation principle. With the large deviation principle $p(A) \sim e^{-tI(A)}$, we can show that the dominant path takes a value of $A(\omega)$ close to

$$A^*(h) \equiv \underset{A}{\text{Argmin}}\,[I(A) - hA]. \tag{2.39}$$

Then, the probability taking $A^*(h)$ is exponentially small:

$$p(A^*) \sim e^{-tI(A^*(h))} \tag{2.40}$$

with $I(A^*(h)) \neq 0$. This means that we need the rare events characterised by a large deviation principle, in order to obtain the information of the large deviation principle itself. We thus conclude that the direct application of (2.38) is not useful. We need some ideas to overcome this difficulty.

2.3.3 Renormalisation of Rare-Eventness via Measurements and Feedbacks

Now, we explain the key idea of our method. It is composed of two parts: Firstly, we consider small values of h in (2.40), more precisely, sufficiently small so that it satisfies

$$t_a I(A^*(h)) = O(1), \tag{2.41}$$

where t_a is the correlation time of $\alpha(\boldsymbol{n} \to \boldsymbol{n}')$. Then, we can easily say that the dominant contribution to obtain $\langle e^{htA(\omega)} \rangle_{\boldsymbol{n}}$ is not rare, which is, in other words, $\langle e^{htA(\omega)} \rangle_{\boldsymbol{n}}$ is measurable for sufficiently small h. Secondly, the exponentially biased measure $P(\omega; h|\boldsymbol{n}_0)$ has a renormalisable structure with respect to the biasing with exponential function. That is,

$$P(\omega; h + h'|\boldsymbol{n}_0) \propto P(\omega; h|\boldsymbol{n}_0) e^{h'tA(\omega)}. \tag{2.42}$$

These two ideas lead to a method of the rare event sampling method. First, for sufficiently small h, we measure $\langle e^{htA(\omega)} \rangle_{\boldsymbol{n}}$. Then, by using the obtained result, we modify the transition rate to create the (probability preserving) system corresponding to the biased measure $P(\omega; h|\boldsymbol{n}_0)$. Then, again in the new system, we measure $\langle e^{htA(\omega)} \rangle_{\boldsymbol{n}}$ for sufficiently small h. Thanks to the property (2.42), the obtained result will lead to the next system corresponding to the biased ensemble $P(\omega; 2h|\boldsymbol{n}_0)$. In this method, just with measuring $e^{htA(\omega)}$, we could reach the biased ensemble $P(\omega; 2h|\boldsymbol{n}_0)$. This is the important property in the rare events of the large deviation principle in time-series statistics. This property can be phrased as follows: *Rare events characterised by the large deviation principle of time-averaged quantity is additive in a sense of measurement. The rare-eventness can be renormalised during the measurements through a feedback (or a modification) to the system.* See Fig. 2.2 for the schematic diagram of this structure.

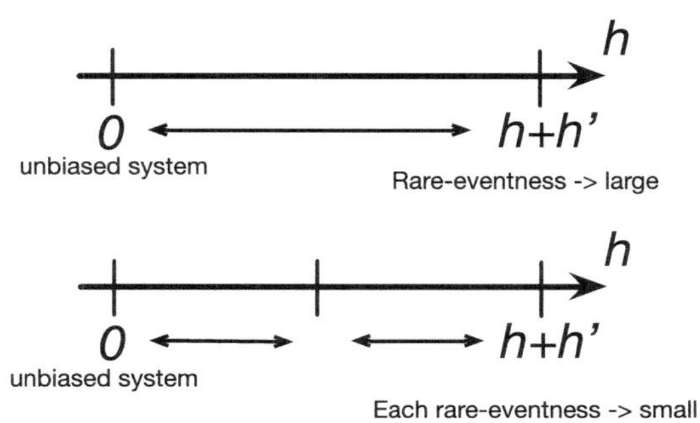

Fig. 2.2 The schematic diagram for renormalisability of rare-eventness

2.3.4 Rare-Event Sampling Method Constituted of an Iterative Measurement-and-Feedback Procedure

By using the property explained above, we propose a method for rare-events sampling method constituted of an iterative measurement-and-feedback procedure.

First we set a measurement time t to be much larger than the correlation time of $\alpha(\boldsymbol{n} \to \boldsymbol{n}')$, t_α. Then, we define a small increment δh from the condition (2.41), or more precisely,

$$t_a \delta h^2 \sigma = O(1), \tag{2.43}$$

where σ is a scaled variance of $A(\omega)$ defined as

$$\sigma = \lim_{t\to\infty} t \left\langle A(\omega)^2 - (\langle A(\omega)\rangle)^2 \right\rangle. \tag{2.44}$$

With this δh, the procedure is defined as follows:

1. For the first step, we measure $\left\langle e^{\tau \delta h A(\omega)} \right\rangle_{\boldsymbol{n}}$ as a function of $\boldsymbol{n}$ in the original system. Here, we remind us that the measurement is not hard because of the condition (2.43).
2. Then, depending on the value of $\left\langle e^{\tau \delta h A(\tau)} \right\rangle_{\boldsymbol{n}}$, we modify the transition rate to
$$w^{\delta h}(\boldsymbol{n} \to \boldsymbol{n}') = w(\boldsymbol{n} \to \boldsymbol{n}') e^{\delta h \alpha(\boldsymbol{n}\to\boldsymbol{n}')} \frac{\left\langle e^{\tau \delta h A(\omega)} \right\rangle_{\boldsymbol{n}'}}{\left\langle e^{\tau \delta h A(\omega)} \right\rangle_{\boldsymbol{n}}}. \tag{2.45}$$
3. Next, in the created modified system, we measure the expected value of the same quantity $e^{\tau \delta h A(\omega)}$. We denote the obtained expected value by $\left\langle e^{\tau \delta h A(\omega)} \right\rangle_{\boldsymbol{n}}^{\delta h}$.
4. Again, we define the second modified transition rate as
$$w^{2\delta h}(\boldsymbol{n} \to \boldsymbol{n}') = w^{\delta h}(\boldsymbol{n} \to \boldsymbol{n}') e^{\delta h \alpha(\boldsymbol{n}\to\boldsymbol{n}')} \frac{\left\langle e^{\tau \delta h A(\omega)} \right\rangle_{\boldsymbol{n}'}^{\delta h}}{\left\langle e^{\tau \delta h A(\omega)} \right\rangle_{\boldsymbol{n}}^{\delta h}}. \tag{2.46}$$
5. We iterate this procedure for many times. Then, we obtain a set of transition rates
$$w^{l\delta h}(\boldsymbol{n} \to \boldsymbol{n}') = w(\boldsymbol{n} \to \boldsymbol{n}') e^{l\delta h \alpha(\boldsymbol{n}\to\boldsymbol{n}')} \prod_{k=0}^{l-1} \frac{\left\langle e^{\tau \delta h A(\omega)} \right\rangle_{\boldsymbol{n}'}^{k\delta h}}{\left\langle e^{\tau \delta h A(\omega)} \right\rangle_{\boldsymbol{n}}^{k\delta h}} \tag{2.47}$$
with $l = 0, 1, 2, \ldots$.
6. Our computational method is based on the following formula. We denote by $\langle f \rangle^h$ the expected value of time-extensive quantities $f(\omega)$ in the system with the modified transition rate w^h ($h = 0, \delta h, 2\delta h, \ldots$). Then, $\langle f \rangle^h$ equal to the expected values by biased ensemble $P(\omega; h)$. That is,

$$\langle f(\omega)\rangle^h \simeq \frac{\left\langle f(\omega)e^{h\tau A(\omega)}\right\rangle}{\left\langle e^{h\tau A(\omega)}\right\rangle}. \tag{2.48}$$

Here and hereafter in this chapter, $\simeq$ represents the asymptotic equality when $\tau \gg \tau_\alpha$.

7. From the formula, we obtain the expected value of any quantity in biased ensemble. For example, for the large deviation function of $A(\omega)$, by combining (2.48) with (2.15), we reaches a formula

$$I(A) = \max_h \left[hA - \sum_{k=0}^{N-1} \langle f(\omega)\rangle^{\tilde{h}} \delta h \right] + O(\delta h^2) \tag{2.49}$$

with $h = N\delta h$. We write this formula as

$$I(A) = \max_h \left[hA - \int_0^h d\tilde{h} \, \langle f(\omega)\rangle^{\tilde{h}} \right], \tag{2.50}$$

which shows the correspondence to the formula (2.18) in equilibrium statistical mechanics.

We showed the basic strategy to derive (2.48) in the previous subsection. For a mathematically rigorous derivation, see Appendix A.4.

2.4 Applications

As a demonstration, we apply our method to non-equilibrium many-body lattice gas models. The first example is an asymmetric simple exclusion process (ASEP) with non equilibrium open boundary conditions, and the second one is Fredrickson–Andersen (FA) model, which is one of kinetically constrained models.

2.4.1 Effective Descriptions of Exponential Family

Before going to the demonstration, we here introduce a strategy to approach many body systems. That is, *effective descriptions of the exponential family*.

The rare trajectories most contributing to $G(h)$ are generated by the modified transition rate $w^h(\boldsymbol{n} \to \boldsymbol{n}')$ in (2.47). Then, the modification rate $\prod_{k=0}^{l-1} \left\langle e^{\tau\delta h A(\omega)}\right\rangle_{\boldsymbol{n}}^{k\delta h}$ in (2.47) is a function of the configuration of the system. Thus, it may be difficult to apply the method to many-body systems, because the degree of the freedom in them exponentially increases and the computation time for obtaining $w^h(\boldsymbol{n} \to \boldsymbol{n}')$ does as well.

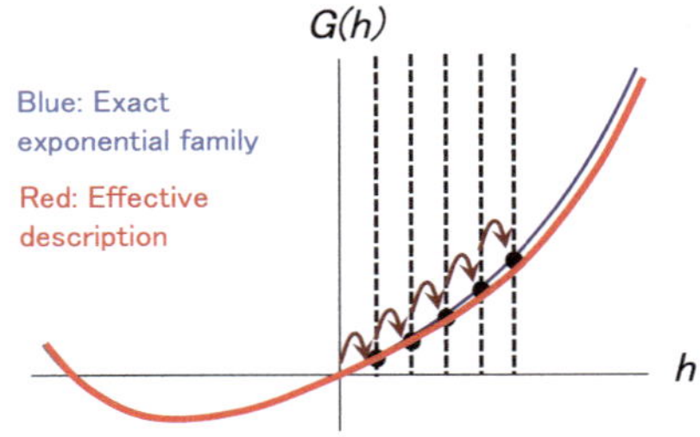

Fig. 2.3 Schematic picture explaining the effective description of exponential family

However, we expect that there are many physical examples that allow us to use the *effective description of the exponential family*. It is to introduce an effective transition rate with K unknown parameters for each value of h, where these unknown parameters are determined by employing (2.47). We then assume that these effective descriptions describe very well the statistical property of the rare trajectories. See Fig. 2.3 for the explanation of this effective description. The system that we are going to analyse in this section has indeed these effective descriptions.

2.4.2 Asymmetric Simple Exclusion Process (ASEP)

Definition of the Model

Let us consider a one dimensional lattice of size L with open boundary conditions. Each site accommodates one particle at most. The configuration of the particles is denoted by $\boldsymbol{n} \equiv (n_i)_{i=1}^{L}$, where n_i takes a value of 1 (occupied) or 0 (empty). The transition rate $w(\boldsymbol{n} \to \boldsymbol{n}')$ is defined as follows: For a configuration $\boldsymbol{n} = (n_1, n_2, n_3, \dots, n_i, n_{i+1}, \dots, n_L)$, we define an exchange operator $F_{i,i+1}$ as

$$F_{i,i+1}\boldsymbol{n} = (n_1, n_2, n_3, \dots, n_{i+1}, n_i, \dots, n_L). \tag{2.51}$$

Also we define a removing, or filling operator F_1 and F_L for the boundaries as

$$F_1 = (1 - n_1, n_2, n_3, \dots, n_i, n_{i+1}, \dots, n_L) \tag{2.52}$$

and

$$F_L = (n_1, n_2, n_3, \dots, n_i, n_{i+1}, \dots, 1 - n_L). \tag{2.53}$$

Then, by using these operators, we define $w(\boldsymbol{n} \to \boldsymbol{n}')$ as

$$w(\boldsymbol{n} \to F_{i,i+1}\boldsymbol{n}) = \delta_{n_i,0}\delta_{n_{i+1},1}q + \delta_{n_i,1}\delta_{n_{i+1},0}, \tag{2.54}$$

$$w(\boldsymbol{n} \to F_1\boldsymbol{n}) = \delta_{n_1,0}\alpha + \delta_{n_1,1}\gamma, \tag{2.55}$$

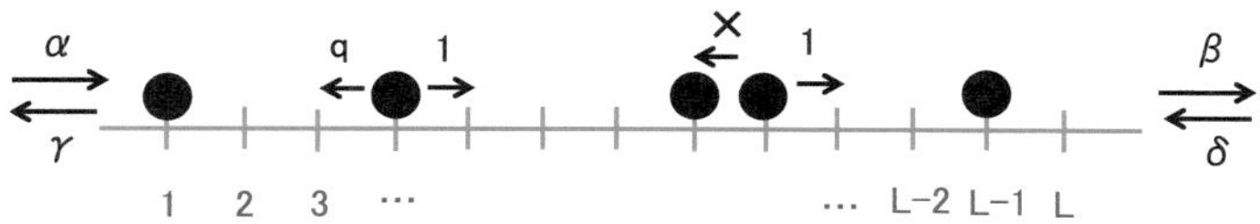

Fig. 2.4 Schematic picture representing the definition of ASEP

$$w(\boldsymbol{n} \to F_L\boldsymbol{n}) = \delta_{n_L,0}\delta + \delta_{n_L,1}\beta, \tag{2.56}$$

and

$$w(\boldsymbol{n} \to \boldsymbol{n}') = 0 \tag{2.57}$$

for any other transition that cannot be expressed by using the operators $F_{i,i+1}$, F_1, and F_L. Equation (2.54) means that a particle moves to the left empty site with a rate q and to the right empty site with a rate 1, when the target site is not occupied. Equation (2.55) and (2.56) represents the injection and the remove of a particle: A particle is injected into the boundary site $i = 1$ ($i = L$) with a rate α (δ) and the particle at the boundary site $i = 1$ ($i = L$) is removed with a rate γ (β). See Fig. 2.4 for the schematic picture to explain these transitions. This model is called ASEP and has been studied as a cornerstone of non-equilibrium physics. See the introduction of Ref. [17] and also Ref. [35] for the review.

In this model, we study the fluctuation of time-averaged bulk current. We first define an instantaneous current at ith site as $j_i(\boldsymbol{n} \to \boldsymbol{n}') = \pm 1$, which takes the value 1 (or -1) when a particle moves from i to $i+1$ ($i+1$ to i). By using this instantaneous current, we then define the bulk current as

$$\alpha(\boldsymbol{n} \to \boldsymbol{n}') = \frac{1}{L-1}\sum_{i=1}^{L-1} j_i(\boldsymbol{n} \to \boldsymbol{n}'). \tag{2.58}$$

The time-averaged bulk current $A(\omega)$ is defined as the time-averaged quantity of this $\alpha(\boldsymbol{n} \to \boldsymbol{n}')$. See Sect. 2.2.2 for the definition or $A(\omega)$.

Numerical Check of Our Formulation

We first verify our formulation numerically. On one hand, we evaluate

$$\prod_{k=0}^{l-1} \left\langle e^{\tau\delta h A(\omega)} \right\rangle_{\boldsymbol{n}}^{k\delta h} \tag{2.59}$$

for a integer l, which is the modification factor appeared in (2.47). On the other hand, with a Monte Carlo simulation, we evaluate the left-eigenvector $\phi(\boldsymbol{n})$ corresponding to the largest eigenvalue of $L^h_{\boldsymbol{n}',\boldsymbol{n}}$ defined in (2.25). According to our formulation, (2.59) is proportional to $\phi(\boldsymbol{n})$, which can be seen by comparing (2.22) with (2.47). See Fig. 2.5 for the examples of the obtained results.

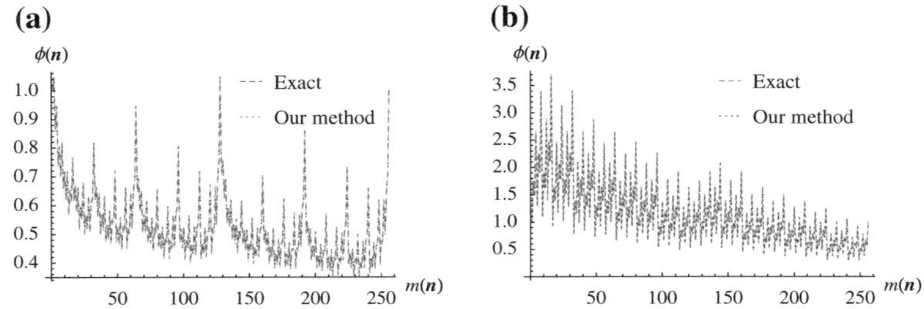

Fig. 2.5 Numerical check of our formulation for the open boundary ASEP. We set $q = 0.5$, $L = 8$, $\alpha = 0.8$, $\beta = 0.8$, $\gamma = 0.2$, and $\delta = 0.2$. We perform our computational method on Monte Carlo simulations with setting $l = 100$ and 300 with $\delta h = -0.02$. The obtained $\prod_{k=0}^{l-1} \langle e^{\tau \delta h A(\omega)} \rangle_{\boldsymbol{n}}^{k\delta h}$ are plotted on **a** (for $l = 100$) and **b** (for $l = 300$) labeled as our method. In the figure, the x axis represents $m(\boldsymbol{n}) = \sum_{i=0}^{L-1} n_{L-i} 2^i$, which is the decimal value of the binary number $\boldsymbol{n}$. On the same figures, we also plot $\phi(\boldsymbol{n})$ that is the left eigenvector corresponding to the largest eigenvalue of $L^h_{\boldsymbol{n}',\boldsymbol{n}}$ defined in (2.25) for $h = -2$ and $h = -6$ (Exact). In these two figures, we can see the coincidence very well between $\prod_{k=0}^{l-1} \langle e^{\tau \delta h A(\omega)} \rangle_{\boldsymbol{n}}^{k\delta h}$ and $\phi(\boldsymbol{n})$. Reprinted with permission from Ref. [27]. Copyright 2014 by American Physical Society

The Effective Description

Next, we study an effective description of the exponential family. First, we define an effective transition rate $w^h_{\rm eff}(\boldsymbol{n} \to \boldsymbol{n}')$ with $L + 1$ unknown parameters $(\psi_{h,i})_{i=0}^{L}$ as

$$w^h_{\rm eff}(\boldsymbol{n} \to F_{i,i+1}\boldsymbol{n}) \equiv w(\boldsymbol{n} \to F_{i,i+1}\boldsymbol{n}) e^{(n_i - n_{i+1})h/(L-1)} \left(\frac{\psi_{h,i+1}}{\psi_{h,i}} \right)^{n_i - n_{i+1}}. \quad (2.60)$$

For the left and right boundary transitions, we also define

$$w^h_{\rm eff}(\boldsymbol{n} \to F_1\boldsymbol{n}) \equiv w(\boldsymbol{n} \to F_1\boldsymbol{n}) (\psi_{h,1}/\psi_{h,0})^{1-2n_1} \quad (2.61)$$

and

$$w^h_{\rm eff}(\boldsymbol{n} \to F_L\boldsymbol{n}) \equiv w(\boldsymbol{n} \to F_L\boldsymbol{n}) (\psi_{h,L}/\psi_{h,0})^{1-2n_L}. \quad (2.62)$$

Here, we note that this new transition rate corresponds to an ASEP that has a spatially varying transition rate as shown in Fig. 2.6. Furthermore, in a sense of local detailed balance condition [1], the effective transition rate represents the system, where a one-body external potential is applied.

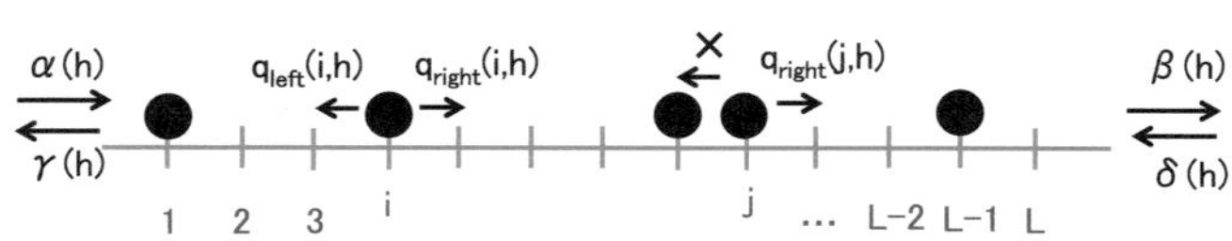

Fig. 2.6 An example of an effective description for ASEP. In this figure, each transition rate is given as follows: $q_{\rm left}(i, h) = qe^{-h/(L-1)}\psi_{h,i-1}/\psi_{h,i}$, $q_{\rm right}(i, h) = qe^{h/(L-1)}\psi_{h,i+1}/\psi_{h,i}$, $\alpha(h) = \alpha\psi_{h,1}/\psi_{h,0}$, $\gamma(h) = \gamma\psi_{h,0}/\psi_{h,1}$, $\beta(h) = \beta\psi_{h,0}/\psi_{h,L}$ and $\delta(h) = \delta\psi_{h,L}/\psi_{h,0}$

We determine the values of the parameters $(\psi_{h,i})_{i=0}^{L}$ from our computational method as follows:

1. For $l = 0$ ($h = 0$), we have $\psi_{h,i} = 1$ for $i = 0, \ldots, L$.
2. For given $(\psi_{l\delta h,i})_{i=0}^{L}$, we determine the next $(\psi_{(l+1)\delta h,i})_{i=0}^{L}$ from the following procedure: We measure $\langle e^{\tau \delta h A(\omega)} \rangle_{\boldsymbol{n}}^{l\delta h}$ for $L+1$ different configurations $\boldsymbol{n} = \boldsymbol{n}_j$ $(j = 0, 1, 2, \ldots, L)$. Especially, here, we choose $(\boldsymbol{n}_j)_i = \delta_{ij}$ as the simplest choice.
3. Next, by applying (2.47) to the effective transition rate (2.60), we obtain

$$\psi_{(l+1)\delta h,i} = \psi_{l\delta h,i} \langle e^{\tau \delta h A(\omega)} \rangle_{\boldsymbol{n}_i}^{l\delta h} \tag{2.63}$$

   for $i = 0, \ldots, L$. Thus, we obtain the next parameters $(\psi_{(l+1)\delta h,i})_{i=0}^{L}$ from $(\psi_{l\delta h,i})_{i=0}^{L}$.
4. By iterating this procedure, we obtain the effective description of the exponential family.

Now, we apply our computational method to Monte-Carlo simulations. We calculated $G(h)$, then plot it in Fig. 2.7. On the same figure, for the comparison, we also plot the largest eigenvalue of $L_{\boldsymbol{n},\boldsymbol{n}'}^{h}$, K, because it is equal to $G(h)$ as shown in (2.31). For further comparison, we also plot the truncated cumulant expansions up to the second order: $G_2(h) = hg_1 + h^2 g_2$ and the fourth order: $G_4(h) = hg_1 + h^2 g_2 + h^3 g_3 + h^4 g_4$, where the coefficients g_i are defined as $(1/i!)\partial^i G(h)/\partial h^i|_{h=0}$. These coefficients are calculated from the exact formula in Refs. [17, 19]. Even though one can see a small deviation between our result (red dotted line) and the exact result (green dashed line) around $h = -7$, the accuracy of our one is considerably better than the one for the truncated cumulant expansions (blue and yellow solid lines). We thus claim that *rare fluctuations of the ASEP in a sense of the large deviation of the current*

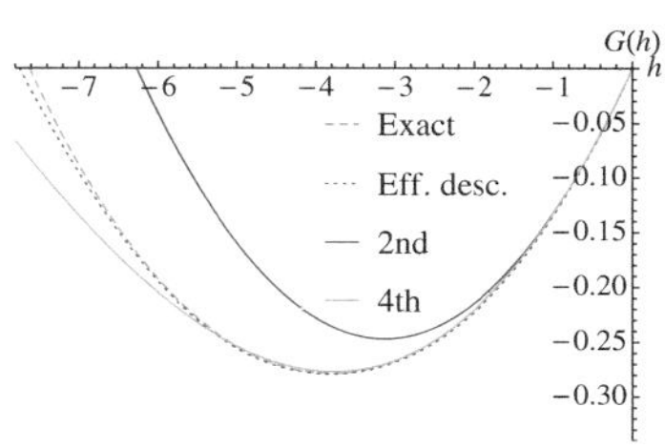

Fig. 2.7 $G(h)$ in the open boundary ASEP obtained from our effective description. We set $q = 0.5$, $L = 8$, $\alpha = 0.8$, $\beta = 0.8$, $\gamma = 0.2$, and $\delta = 0.2$. Following the procedure for the effective description described in the text, we perform Monte Carlo simulations with fixed $\delta h = -0.02$. The result is labeled as Eff. desc. For comparison, we also plot the largest eigenvalue of $L_{\boldsymbol{n},\boldsymbol{n}'}^{h}$ (Exact), the truncated cumulant expansions up to the second order: $G_2(h) = hg_1 + h^2 g_2$ (2nd) and the fourth order: $G_4(h) = hg_1 + h^2 g_2 + h^3 g_3 + h^4 g_4$ (4th), where the coefficients g_i are defined as $(1/i!)\partial^i G(h)/\partial h^i|_{h=0}$. These coefficients are calculated from the exact formula in Refs. [17, 19]. Reprinted with permission from Ref. [27]. Copyright 2014 by American Physical Society

with this parameter set are well described by the effective transition rate (2.60).We expect that there are some mathematical formulas related to this observation.For this, we mention a variational principle determining the large deviation function of the current in lattice gas models in thermodynamic limit (infinite size limit), which was proposed in [7, 8]. In the paper, Bodineau and Derrida derived the formula from a phenomenology called *an additivity principle.* If we restrict ourselves to the system of SSEP or (WASEP), this variational principle, on the other side, can be derived from the general variational principle given as (2.30). In the derivation, we assume the effective transition rate in thermodynamic limit [36]. We need further investigation for clarifying the applicability of this effective-description approach upon general lattice gas models.

2.4.3 Fredrickson–Andersen (FA) Model

Definition of the Model

Next, we consider a Fredrickson–Andersen (FA) model [37, 38]. This is an example of kinetically constrained models (KCMs), which has been studied for understanding the glassy features from the dynamical aspect of the system. We define an occupation variable $n_i = 1$ or 0 on each site of a one-dimensional lattice. The size of the lattice is L, and the boundary condition is periodic. For a configuration $\boldsymbol{n} = (n_1, \ldots, n_L)$, we define a flipping operator C_i as

$$C_i \boldsymbol{n} = (n_1, \ldots, 1 - n_i, \ldots, n_L). \tag{2.64}$$

Then, from a configuration $\boldsymbol{n}$ to $C_i \boldsymbol{n}$, we define the corresponding transition rate as

$$w(\boldsymbol{n} \to C_i \boldsymbol{n}) = [(1 - c)n_i + c(1 - n_i)] f_i(\boldsymbol{n}), \tag{2.65}$$

with the definition of $f_i(\boldsymbol{n})$ as

$$f_i(\boldsymbol{n}) \equiv n_{i-1} + n_{i+1}. \tag{2.66}$$

See Fig. 2.8 for the schematic picture explaining the definition of this FA model. The transition rate is a product of two parts. The one is a part without any interaction $[(1-c)n_i + c(1-n_i)]$, and the other one is a part for the kinetically constraint $f_i(\boldsymbol{n})$. Only the first part takes responsibility for the stationary state. Indeed, if we consider the detailed balance condition, $f_i(\boldsymbol{n})$ is canceled out. Then the stationary probability $p(\boldsymbol{n})$ is just determined from this non interacting part $[(1-c)n_i + c(1-n_i)]$, which leads to

$$p(\boldsymbol{n}) = \prod_{i=1}^{L} [cn_i + (1 - c)(1 - n_i)]. \tag{2.67}$$

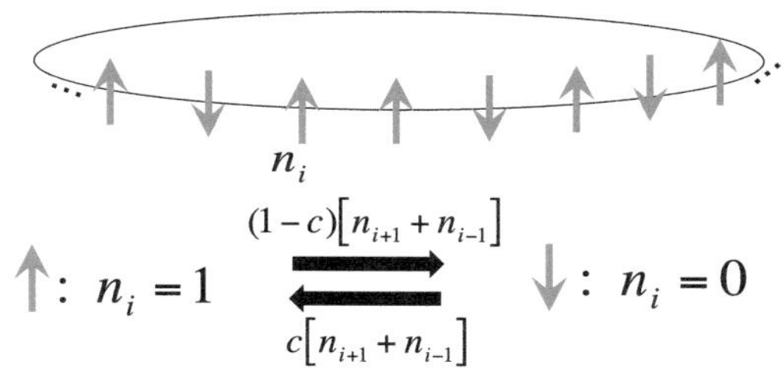

Fig. 2.8 Schematic picture explaining FA model

The second part $f_i(\boldsymbol{n})$ actually represents a kinetic constraint. For example, we look at the configuration, where ith site is surrounded by unoccupied site. In this case, the flipping is indeed blocked due to $f_i(\boldsymbol{n})$. Although the stationary state is trivial, the system shows the same features as the one in glassy systems, due to this kinetic constraint. See Ref. [15, 39] for this review. Recently, for studying these features, the approach using a large deviation principle gathered attention. In 2007, Garrahan, Jack, Lecomte, Pitard, van Duijvendijk and van Wijland considered a dynamical activity defined as

$$\alpha(\boldsymbol{n} \to \boldsymbol{n}') = 1, \tag{2.68}$$

which represents how often the state of the system changes. Then, in several KCMs, they numerically calculated the cumulant generating function of this time-averaged activity, and found the singularity in it in $L \to \infty$ [15, 16, 40]. This represents a dynamical phase transition of the system, which is believed to be related to dynamical heterogeneities. After the finding, the finite size effect of the singularity has been studied by Bodineau, Lecomte and Toninelli [41, 42]. Since the system that we can simulate is always finite, the study to extract the property of dynamical phase transition from finite-size systems is important. In this section, we approach to this problem with our formulation, especially, by looking at the effective description of the system.

The Effective Description

First, we define the effective transition rate as

$$w^h_{\text{eff}}(\boldsymbol{n} \to C_i\boldsymbol{n}) \equiv w(\boldsymbol{n} \to C_i\boldsymbol{n})e^h[C^h((n_{i\pm j})^r_{j=1})]^{1-2n_i}, \tag{2.69}$$

where r is a truncating number of the interaction range and the function $C^h((n_{i\pm j})^r_{j=1})$ is an unknown function of local variables. See Fig. 2.9 for the schematic picture explaining this effective transition rate. This is defined for investigating how much the long-rang interactions can affect the dynamical phase transition. The transition rate becomes more accurate as r increases up to $r \simeq L/2$. For fixed r, like the application to ASEP in the previous subsection, we determine the function $C^h((n_{i\pm j})^r_{j=1})$ as follows:

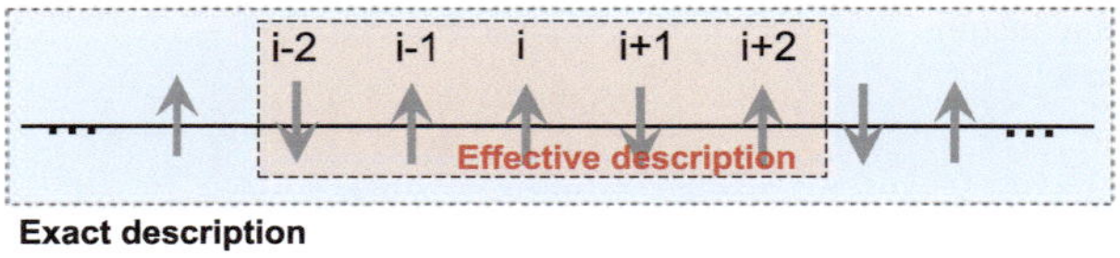

Fig. 2.9 Schematic picture explaining an example of the effective description in FA model

1. For $l = 0$ ($h = 0$), we have $C^h((n_{i\pm j})_{j=1}^r) = 1$.
2. For given $C^{l\delta h}((n_{i\pm j})_{j=1}^r)$, we determine the next $C^{(l+1)\delta h}((n_{i\pm j})_{j=1}^r)$ with the following procedure: We measure $\left\langle e^{\tau\delta h A(\omega)}\right\rangle_{\boldsymbol{n}}^{l\delta h}$ for 2^{2r+1} different configurations $\boldsymbol{n} = \boldsymbol{n}_j$ ($j = 1, 2, \ldots, 2^{2r+1}$). Here, we choose $\boldsymbol{n}_j = (0, n_2, \ldots, n_{r+1}, 0, \ldots, 0, n_{L-r+1}, \ldots n_L)$ with $n_i = 1$ or 0 ($i = 2, \ldots, r+1, L-r+1 \ldots, L$) and $\boldsymbol{n}_j = (1, n_2, \ldots, n_{r+1}, 0, \ldots, 0, n_{L-r+1}, \ldots n_L)$ with $n_i = 1$ or 0 ($i = 2, \ldots, r+1, L-r+1 \ldots, L$). Because the system has translational invariance, we can regard the site $i = 1$ as the centre of the system without loss of generality.
3. Next, by applying (2.47) to the effective transition rate (2.60), we obtain
$$C^{(l+1)\delta h}((n_{1\pm j})_{j=1}^r) = C^{l\delta h}((n_{1\pm j})_{j=1}^r)\frac{\left\langle e^{\tau\delta h A(\omega)}\right\rangle_{C_1\boldsymbol{n}}^{l\delta h}}{\left\langle e^{\tau\delta h A(\omega)}\right\rangle_{\boldsymbol{n}}^{l\delta h}} \tag{2.70}$$
for $i = 1, \ldots, 2^{2r+1}$. Thus, we obtain the next parameters $C^{(l+1)\delta h}((n_{1\pm j})_{j=1}^r)$ from $C^{l\delta h}((n_{1\pm j})_{j=1}^r)$.
4. By iterating this procedure, we obtain the effective description of the exponential family.

First, we check the validity of our formulation by launching Monte Carlo simulations for small system sizes. We diagonalise the matrix $L_{\boldsymbol{n}',\boldsymbol{n}}^h$ defined in (2.25), then we calculate
$$\phi(C_1\boldsymbol{n})/\phi(\boldsymbol{n}) \tag{2.71}$$
with $\boldsymbol{n} = (0, n_2, \ldots, n_{r+1}, 0, \ldots, 0, n_{L-r+1}, \ldots n_L)$, which corresponds to $C^h((n_{1\pm j})_{j=1}^r)$ in our formulation. The examples of this result are shown in Fig. 2.10a.

Next, we investigate the long-range nature of the dynamical phase transition. We fix relatively large values of L, where the direct diagonalization of the matrix $L_{\boldsymbol{n}',\boldsymbol{n}}^h$ is too demanding. Then, we launch the Monte Carlo simulations and obtain $G(h)$ for several values of r. On the other hand, we apply the population dynamics method to the same system, which is a numerical technique to calculate large deviation functions [43, 44]. We plot all the obtained results in Fig. 2.10b. In the figure, we can see that the curves obtained $r = 3, 4$ shows the convergence, especially for the region $h < 0$. Since this part $h < 0$ takes responsibility for the dynamical phase transition explained above [41, 42], this result suggests that the long-range interactions for the modified transition rate is not relevant to it. We mention that the long-range nature of

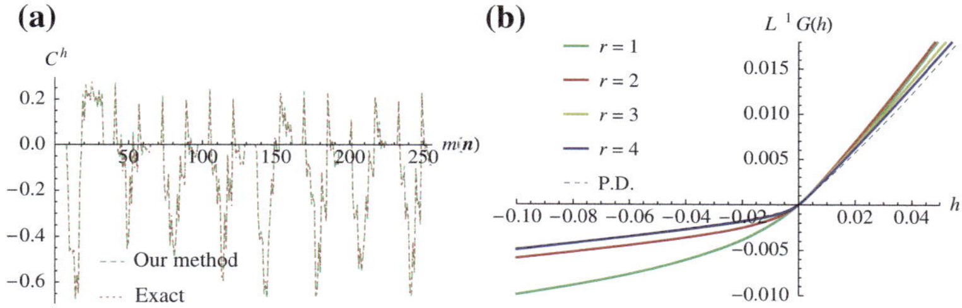

Fig. 2.10 Statistical properties of an activity for the FA model with $c = 0.3$. **a** We set $L = 10, r = 4$, $l = 40$ and $\delta h = -0.0025$, then we perform our formulation. We plot obtained $C^{l\delta h}((n_{1+j})_{j=1}^{r})$ (Our method), where the x axis represents $m(\boldsymbol{n}) = \sum_{i=0}^{r-1} n_{L-i}2^i + \sum_{i=r}^{2r-1} n_{2r+1-i}2^i$. On the same figure, we also plot $\phi(C_1\boldsymbol{n})/\phi(\boldsymbol{n})$ with $\boldsymbol{n} = (0, n_2, \ldots, n_{r+1}, 0, \ldots, 0, n_{L-r+1}, \ldots n_L)$ obtained from the left eigenvector corresponding to the largest eigenvalue of $L^h_{\boldsymbol{n},\boldsymbol{n}'}$ for $L = 10$ with $h = -0.1$ (Exact). **b** We set $L = 30$ and $\delta h = -0.0025$, and we perform our formulation for several r. The obtained $G(h)/L$ are plotted on the figure. At the same time, we also perform the population dynamics method [43, 44] for obtaining $G(h)/L$. The obtained result is plotted as a black dashed line (P. D.) in the figure. Reprinted with permission from Ref. [27]. Copyright 2014 by American Physical Society

the effective interactions has also been studied very recently in Ref. [45] for the East model. They solved analytically a variational principle that characterises modified systems (which is the same as the one that we explained in Sect. 2.2.5). Then, they focused on the part $h > 0$ and concluded that the long range interaction could not be negligible. This result is not contradictory to ours, because they focused on the different region of h, and also, they measured different quantities from ours. Indeed, in Fig. 2.10b for $h > 0$, we can see a small difference between results with effective interactions and the one with population dynamics.

In this system, for investigating the singular behaviour of $G(h)$ more precisely, a scaled biasing parameter $\tilde{h} \equiv hL$ has been introduced by Bodineau, Lecomte, and Toninelli [41, 42]. They proved that

$$\tilde{G}(\tilde{h}) = G(\tilde{h}/L) \tag{2.72}$$

is not an analytic function in the limit $L \to \infty$. However, the nature of the singularity, for example how the singularity arrises as the system size becomes larger, has not been understood yet. The problem was in the numerical study of it because the population dynamics method does not exhibit good convergence of $\tilde{G}(\tilde{h})$ for relatively large values of L [41]. In Appendix A.5, we show that our method can also be applied for obtaining the reliable L dependence of $\tilde{G}(\tilde{h})$ even in this situation.

2.5 Conclusion

In this chapter, we studied the phenomenological structure for the large deviation principle in time-series statistics. By using this structure, we proposed a rare-event sampling method composed of a measurement and feedback, which produced a set of transition rates that had the same statistical properties as those in the biased ensemble of large deviation statistics. For applying the method to spatially extended many-body systems, where the number of degrees of freedom increased exponentially, we also proposed a method to construct an effective description of the biased ensemble. The example of the effective description is as follows: For the case of ASEP, spatially varying one-body potentials instead of many-body potentials can be this effective description. For FA model, the finite-size effect of dynamical phase transition appearing in $G(h)$ is well described by the effective description without long-range interactions.

Here, we mention a future possibility related to this effective description. In order to get a good effective description, physical intuition and some efforts with trial-and-error are needed. But once after we get a good effective description, the computation time of large deviation statistics will be shortened very much. What we need to know now is the theory to determine such effective descriptions automatically if a system is given. For constructing this theory, we need many examples of effective descriptions in many systems. We start with a simple problem such as heat conduction, then increase the complexity of the problem gradually. One of the challenging goals to achieve is to find an effective description of fully-developed turbulence. (See Conclusion for the detail.) Since there is an important problem related to rare-events in turbulence, such a description, if it is found, will certainly promote the understanding of turbulence, especially from a viewpoint of rare-event sampling application to real experiments.

References

1. G.E. Crooks, Phys. Rev. E **60**, 2721 (1999)
2. D.J. Evans, E.G.D. Cohen, G.P. Morriss, Phys. Rev. Lett. **71**, 2401 (1993)
3. G. Gallavotti, E.G.D. Cohen, Phys. Rev. Lett. **74**, 2694 (1995)
4. J. Kurchan, J. Phys. A **31**, 3719 (1998)
5. J.L. Lebowitz, H. Spohn, J. Stat. Phys. **95**, 333 (1999)
6. C. Maes, J. Stat. Phys. **95**, 367 (1999)
7. T. Bodineau, B. Derrida, Phys. Rev. Lett. **92**, 180601 (2004)
8. T. Bodineau, B. Derrida, J. Stat. Phys. **123**, 277 (2006)
9. T. Bodineau, B. Derrida, C. R. Phys. **8**, 540 (2007)
10. B. Derrida, J. Stat. Mech. P07023 (2007)
11. B. Derrida, Phys. Rep. **301**, 65 (1998)
12. B. Derrida, J.L. Lebowitz, Phys. Rev. Lett. **80**, 209 (1998)
13. L. Bertini, A. De Sole, D. Gabrielli, G. Jona-Lasinio, C. Landim, Phys. Rev. Lett. **94**, 030601 (2005)
14. L. Bertini, A. De Sole, D. Gabrielli, G. Jona-Lasinio, C. Landim, J. Stat. Phys. **123**, 237 (2006)

15. J.P. Garrahan, P. Sollich, C. Toninelli, arXiv:1009.6113 (2010), which is a chapter in Ref. L. Berthier, G. Biroli, J.-P. Bouchaud, L. Cipelletti, W. van Saarloos (eds.), *Dynamical Heterogeneities in Glasses, Colloids, and Granular Media* (Oxford University Press, Oxford, 2011)
16. J.P. Garrahan, R.L. Jack, V. Lecomte, E. Pitard, K. van Duijvendijk, F. van Wijland, Phys. Rev. Lett. **98**, 195702 (2007)
17. M. Gorissen, A. Lazarescu, K. Mallick, C. Vanderzande, Phys. Rev. Lett. **109**, 170601 (2012)
18. C. Maes, K. Netočný, B. Wynants, Phys. Rev. Lett. **107**, 010601 (2011)
19. A. Lazarescu, K. Mallick, J. Phys. A: Math. Theor. **44**, 315001 (2011)
20. R. Chetrite, H. Touchette (2014). arXiv:1405.5157
21. R. Chetrite, H. Touchette, Phys. Rev. Lett. **111**, 120601 (2013)
22. R.M.L. Evans, Phys. Rev. Lett. **92**, 150601 (2004)
23. R.L. Jack, P. Sollich, Prog. Theor. Phys. Suppl. **184**, 304 (2010)
24. T. Nemoto, S. Sasa, Phys. Rev. E **83**, 030105(R) (2011)
25. T. Nemoto, S. Sasa, Phys. Rev. E **84**, 061113 (2011)
26. S. Sasa, Phys. Scr. **86**, 058514-1-3 (2012)
27. T. Nemoto, S. Sasa, Phys. Rev. Lett. **112**, 090602 (2014)
28. C.W. Gardiner, *Handbook of Stochastic Methods for Physics, Chemistry, and the Natural Sciences* (Springer, Berlin, 1983)
29. H. Touchette, Phys. Rep. **478**, 1 (2009)
30. E. Seneta, *Non-Negative Matrices and Markov Chains*, 2nd edn. (Springer, New York, 2006)
31. M.D. Donsker, S.R. Varadhan, Commun. Pure Appl. Math. **28**, 1 (1975)
32. R. van Zon, E.G.D. Cohen, Phys. Rev. Lett. **91**, 110601 (2003)
33. R. van Zon, E.G.D. Cohen, Phys. Rev. E **69**, 056121 (2004)
34. A. Dembo, O. Zeitouni, *Large Deviations Techniques and Applications* (Springer, New York, 1998)
35. A. Parmeggiani, Physics **5**, 118 (2012)
36. In preparation of submitting paper
37. G.H. Fredrickson, H.C. Andersen, Phys. Rev. Lett. **53**, 1244 (1984)
38. G.H. Fredrickson, H.C. Andersen, J. Chem. Phys. **83**, 5822 (1985)
39. F. Ritort, P. Sollich, Adv. Phys. **52**, 219 (2003)
40. J.P. Garrahan, R.L. Jack, V. Lecomte, E. Pitard, K. van Duijvendijk, F. van Wijland, J. Phys. A **42**, 075007 (2009)
41. T. Bodineau, V. Lecomte, C. Toninelli, J. Stat. Phys. **147**, 1 (2012)
42. T. Bodineau, C. Toninelli, Commun. Math. Phys. **311**, 357 (2012)
43. C. Giardina, J. Kurchan, L. Peliti, Phys. Rev. Lett. **96**, 120603 (2006)
44. C. Giardina, J. Kurchan, V. Lecomte, J. Tailleur, J. Stat. Phys. **145**, 787 (2011)
45. R.L. Jack, P. Sollich, J. Phys. A **47**, 015003 (2014)

Chapter 3
Common Scaling Functions in Dynamical and Quantum Phase Transitions

3.1 Introduction

KCMs (Kinetically constrained models) have a dynamical phase transition when it is biased by the activity of the system [1–3]. For studying the singularity appearing in the dynamical free energy, numerical simulations of KCMs are useful. The main obstacle here is that the singularity appears in infinite system size limit. For finite size systems, which are only the settings that we can consider in numerical simulations, we need some special technique to extract the property of the singularity. This is the purpose of *finite size scaling*, which has been studied by Bodineau, Toninelli, and Lecomte [4, 5]. Until now, several mathematical results have been obtained. However, the understanding is not yet enough to make quantitative connections between the dynamical phase transition and the glassy features such as dynamical heterogeneity. Further studies are needed.

For the finite-size scaling, one important problem, which is specific to thermodynamic formalism in time-series statistics, arrises. It is related to the lack of the direct connection from a biased ensemble to a physical system (or Monte Carlo simulation): Even for finite-size systems, a special value of the biasing field s can be defined as the one giving the local maximum of the second derivative of the cumulant generating function. However, this value is deviated from the origin whenever we consider finite size systems [4, 5]. Only rare trajectories, generated by an exponential biasing, is related to this special point.

Here, we remind us of the *phenomenological structure for the large deviation principle* explained in Chap. 1. With this structure, such rare trajectories are realised in a stationary state of another (modified) system. Until now, some studies that focus on this structure in the dynamical phase transition has been done, for example by Jack, Sollich [6] and by us [7]. The studies are based on some numerical technique, and any analytical expression, which accelerates the understanding of dynamical phase transition, is not obtained yet. Especially, there is a lack of the detailed study on a *simple and solvable example*, like mean-field model of ferromagnet. In this chapter,

T. Nemoto, *Phenomenological Structure for the Large Deviation Principle in Time-Series Statistics*, Springer Theses,
DOI 10.1007/978-981-287-811-3_3

we conduct such a detailed study on a mean field model, which is the simplest KCMs called mean-field FA model. We focus on the Hamiltonian corresponding to the dynamical phase transition point. We find that the Hamiltonian itself is a singular function with infinite system size limit, and that the system is naturally divided into two regions, namely *active and inactive regions* due to that singular point. This is explained in Sect. 3.2.2 in this chapter.

Furthermore, we have some by-products thanks to this formulation. One of them is a scaling function around the dynamical phase transition point. For equilibrium ferromagnetic models, a finite size scaling function was found by Borgs and Koteckeý in 1990 [8, 9]. In their theory, they assumed that the partition function at the coexistence region was written as the sum of each partition function corresponding to each coexisting phase. From this assumption, they derived a universal scaling function for those models. For the case of dynamical phase transition, we ask if the same argument is possible or not. In this chapter, we answer "yes" to this question. But the theory of Borgs and Koteckeý cannot be applied directly. We need another procedure that is intrinsic to the time-series statistics. It is the variational principle for determining the corresponding stationary state (2.29). We explain the detail in Sect. 3.2.3 in this chapter.

Interestingly, the scaling function obtained in the dynamical phase transition can be seen in another type of phase transition. It is quantum phase transition, which is defined in temperature zero limit in quantum systems. Mathematically, it is known that the statistical physics with zero-temperature limit and biased ensemble defined in time-series statistics are equivalent. (We revise this known fact in Sect. 3.3.1.) Thus, a natural question will arise that the property obtained in the dynamical phase transition can be imported to the quantum phase transition. We also answer "yes" to this question and show that transverse mean-field Ising spin has exactly the same scaling function. Until now, in the studies of the quantum phase transitions, several interesting results have been obtained. As an example, an exponentially small energy gap between the ground state and the excited state has been studied by Jörg et al. [10] and Bapst and Semerjian [11], and some quantitative formulas about this gap have been reported. We show that our result is compatible with these previous ones. We apply our formulation to a quantum ferromagnet in order to determine the width of the coexistence region, and show that the obtained formula is equivalent to the one derived by Bapst and Semerjian in Ref. [11], which gives the energy gap explained above at the quantum phase transition point.

Here is the organisation of this chapter. In Sect. 3.2, we analyse a mean-field Fredrickson–Andersen (FA) model. Section 3.2.1 is devoted to preliminaries, where the definition of the model and the dynamical activity are given. The known result of the dynamical phase transition in this simple model is also shown in this subsection. In Sect. 3.2.2, we show the analytical expression of the modified Hamiltonian at the transition point, from which the domain of the state of the system can be naturally divided into two regions, namely, active and inactive phases. In Sect. 3.2.3, we propose an ansatz to obtain the scaling function around the dynamical phase transition, which is valid not only in the dynamical phase transition but also in quantum phase transitions. In Sect. 3.3, as announced before, we analyse the quantum phase transition for a mean-field quantum ferromagnet model. In Sect. 3.3.1, we show the mathematical equivalence between the dynamical phase transition and the quantum phase transition. In the next Sect. 3.3.2, we apply our formulation to this system. We first derive the same scaling functions as the one obtained for KCMs. Then, we derive a formula that has the same expression as the one derived by Bapst and Semerjian in Ref. [11]. In Sect. 3.4, we make a conclusion of this chapter. The discussion in this chapter is based on our published paper [12]

3.2 Finite-Size Structure in Mean-Field FA Model

3.2.1 Preliminaries

Definition of the Model

Let us consider a lattice with site L, where an occupation variables $\boldsymbol{n} = (n_i)_{i=1}^{L}$ is defined on each site. We don't specify the shape of the lattice because we consider mean-field model defined below. The variable n_i takes 0 (unoccupied) or 1 (occupied). And follows a continuos-time-Markov dynamics. (See the Sect. 2.2.1 for some basics about continuous-time Markov dynamics). With parameter c that takes value between 0 and 1, the transition rate $w(\boldsymbol{n} \to \boldsymbol{n}')$ is defined as

$$w(\boldsymbol{n} \to C_i\boldsymbol{n}) = [(1-c)n_i + c(1-n_i)]f_i^{\mathrm{M}}(\boldsymbol{n}), \tag{3.1}$$

where C_i is the spin-flip operator defined in (2.64) and $f_{\mathrm{M}}(\boldsymbol{n})$ is a fully connected kinetic constraint defined as

$$f_i^{\mathrm{M}}(\boldsymbol{n}) \equiv \frac{1}{L}\sum_{i \neq j} n_i. \tag{3.2}$$

As explained in Sect. 2.4.3, this constraint factor $f_i^{\mathrm{M}}(\boldsymbol{n})$ doesn't affect the equilibrium distribution function.

Now we change the variable of the system to more simple quantity. Unlike the 1-FA model introduced in Sect. 2.4.3, in this case, the evolution equation has an exact-closed description. By defining a total spin by

$$n = \sum_i n_i, \tag{3.3}$$

we obtain an alternative continuous time Markov dynamics of n with a transition rate

$$w(n \to n+1) = \sum_i (1-n_i)\frac{c}{L}\sum_{j \neq i} n_j = \frac{1}{-}Lcn(L-n) \tag{3.4}$$

and

$$w(n \to n-1) = \sum_i n_i \frac{1-c}{L} \sum_{j \neq i} n_j = \frac{1}{L}(1-c)n(n-1)\,. \tag{3.5}$$

Note that there are no transitions for reaching $n = 0$. We choose the initial condition as the one that doesn't have any probability of $n = 0$, so that the system never reach $n = 0$ state after that. The domain of n is thus $1 \leq n \leq L$. In this system, the escape rate is written as

$$\lambda(n) = \sum_{n'} w(n \to n') = cn(1 - n/L) + (1-c)n(n-1)/L. \tag{3.6}$$

The transition rate satisfies a detailed balance condition

$$P_{\mathrm{eq}}(n)w(n \to n') = P_{\mathrm{eq}}(n')w(n' \to n) \quad (\forall\, n, n') \tag{3.7}$$

with respect to the equilibrium distribution function $P_{\mathrm{eq}}(n)$

$$P_{\mathrm{eq}}(n) = \frac{L!}{n!(L-n)!} \frac{1}{1-(1-c)^L} c^n (1-c)^{L-n}. \tag{3.8}$$

The expected value of n is

$$\langle n \rangle_{\mathrm{eq}} = cL + O((1-c)^L), \tag{3.9}$$

the variance of n is

$$\left\langle (n - \langle n \rangle)^2 \right\rangle_{\mathrm{eq}} = c(1-c)L + O((1-c)^L). \tag{3.10}$$

Here, we note that the parameter c is interpreted as the mean density of occupied sites in the large size limit from (3.9). Also, from (3.9) and (3.10), we obtain the expected value of the escape rate as

$$\langle \lambda \rangle_{\mathrm{eq}} = 2(1-c)c^2 L + O(1). \tag{3.11}$$

There is a large deviation principle for the probability distribution of the fraction $\rho = n/L$. Indeed, from the distribution function of n (3.8), we obtain an asymptotic expression of $L^{-1} P_{\mathrm{eq}}(L\rho)$

$$L^{-1} P_{\mathrm{eq}}(L\rho) \sim e^{-Lf_{\mathrm{e}}(\rho)} \tag{3.12}$$

with

$$f_{\mathrm{e}}(\rho) = (1-\rho)\log\frac{1-\rho}{1-c} + \rho \log\frac{\rho}{c} \tag{3.13}$$

for large L. The function $f_{\mathrm{e}}(\rho)$ is the large deviation function. Here and hereafter in this chapter, we use the *free energy* terminology when we mention the particle number large deviation function $f_{\mathrm{e}}(\rho)$. As explained below, when we indicate the large deviation function of time-averaged quantity, we refer to them as *dynamical free energy*.

Dynamical Activity

For a given transition $\boldsymbol{n} \to \boldsymbol{n}'$, we define a quantity $\alpha(\boldsymbol{n} \to \boldsymbol{n}')$ as

$$\alpha(\boldsymbol{n} \to \boldsymbol{n}') = 1. \tag{3.14}$$

Because $\alpha(\boldsymbol{n} \to \boldsymbol{n}')$ represents how the system actives, it is called dynamical activity. Then, for a given path ω generated by continuos-time Markov dynamics (See the Sect. 2.2.1, for exapmle), the time-averaged dynamical activity $A(\omega)$ during time interval t is

$$A(\omega) = \frac{N}{t}, \tag{3.15}$$

where N is the total number of jumps. The expected value of $A(\omega)$ in stationary state is equal the one of λ. Indeed,

$$\langle A(\omega)\rangle = \sum_{n,n'} P_{\mathrm{eq}}(n)w(n \to n')\alpha(n \to n') = \sum_{n,n'} P_{\mathrm{eq}}(n)w(n \to n') = \langle\lambda\rangle . \tag{3.16}$$

Thus, like (3.11), it is also calculated as

$$\langle A(\omega)\rangle = 2(1-c)c^2 L + O(1). \tag{3.17}$$

Fluctuation of the Dynamical Activity

Although the expected value of $A(\omega)$ is trivial, higher moment has singular property as shown below. For this, we define the cumulant generating function, or dynamical free energy as

$$G(s) = \frac{1}{L}\lim_{t\to\infty}\frac{1}{t}\log\left\langle \mathrm{e}^{-sK_t}\right\rangle . \tag{3.18}$$

We note that the definition of the biasing parameter is different from the one in Chap. 2, h. The relation between h and s is $s = -h$.

From (2.31), $G(s)$ is calculated as the largest eigenvalue of a matrix $L^s_{n',n}$:

$$L^s_{n',n} \equiv w(n \to n')e^{-s} - \delta_{n,n'}\lambda(n) \tag{3.19}$$

and

$$\sum_{n'} \phi(n')L^s_{n',n} = LG(s)\phi(n). \tag{3.20}$$

The Variational Principle for Determining *G(s)*

Here, we recall the variational principle for determining $G(s)$ given as (2.30). In this case, it is written as

$$LG(s) = \max_{\Delta \widetilde{F}} \sum_{n} \widetilde{P}(n) \left[\tilde{\lambda}(n) - \lambda(n) \right], \tag{3.21}$$

where the distribution function $\widetilde{P}(n)$ and modified escape rate $\tilde{r}(n)$ are defined as

$$\widetilde{P}(n) = \frac{P_{\text{eq}}(n) e^{-\Delta \widetilde{F}(n)}}{\sum_{\tilde{n}} P_{\text{eq}}(\tilde{n}) e^{-\Delta \widetilde{F}(\tilde{n})}} \tag{3.22}$$

and

$$\begin{aligned} \tilde{\lambda}(n) = & \, nc \left(1 - \frac{n}{L}\right) \mathrm{e}^{-\frac{1}{2}[\Delta \widetilde{F}(n+1) - \Delta \widetilde{F}(n)] - s} \\ & + n(1-c) \left(\frac{n}{L} - \frac{1}{L}\right) \mathrm{e}^{-\frac{1}{2}[\Delta \widetilde{F}(n-1) - \Delta \widetilde{F}(n)] - s}. \end{aligned} \tag{3.23}$$

Due to (2.29), which represents that the optimal system corresponds to the biased system characterised by the largest eigenvector, the optimal $\Delta \widetilde{F}(n)$ is connected to $\phi(n)$ as

$$\Delta \widetilde{F}(n) = -2 \log \tilde{\phi}(n), \tag{3.24}$$

In the system that satisfies the detailed balance condition, like the system that we are now considering, the formula (3.21) can be mapped to a well-known variational principle determining the ground state energy in quantum mechanics. See Appendix B.4. Then, that equivalent formula to (3.21) has been used to detect the dynamical phase transition [1–3]. Next, as a preliminary of the dynamical phase transition, we show this argument with a viewpoint of a large deviation principle, and introduce the dynamical phase transition in this system.

Dynamical Phase Transition

Here, we show the dynamical phase transition with the variational principle (3.21). First, we assume a large deviation principle in the optimal modified system:

$$\widetilde{P}(n) \sim \mathrm{e}^{-L\left[f_{\mathrm{e}}(n/L) + \tilde{f}(n/L)\right]} \tag{3.25}$$

for large L, where $\tilde{f}(\rho)$ is an unknown free energy we determine below. This is equivalent to an assumption that the variational free energy $\Delta \widetilde{F}(n)$ scales like $L \tilde{f}(\rho)$ with $\rho = n/L$ for large L. Then with these assumptions, we rewrite (3.23) as

$$\frac{\tilde{r}(L\rho)}{L} = \mathrm{e}^{-s} \left[\rho c \, (1-\rho) \, \mathrm{e}^{-\frac{1}{2} \tilde{f}'(\rho)} + \rho^2 (1-c) \mathrm{e}^{\frac{1}{2} \tilde{f}'(\rho)} \right] + O(1/L) \tag{3.26}$$

where $\tilde{f}'(\rho) = \frac{\partial \tilde{f}(\rho)}{\partial \rho}$. By using this expression in (3.21), we obtain the leading term of it as

$$G(s) = \max_{\tilde{f}>0} \frac{\int d\rho \left\{ e^{-s}\left[\rho c(1-\rho)e^{-\frac{1}{2}\tilde{f}'(\rho)} + \rho^2(1-c)e^{\frac{1}{2}\tilde{f}'(\rho)}\right] - r(L\rho)/L \right\} e^{-L[f_e(\rho)+\tilde{f}(\rho)]}}{\int d\rho \, e^{-L[f_e(\rho)+\tilde{f}(\rho)]}} \tag{3.27}$$

Then, we perform the saddle point approximation in the integrals over ρ with an assumption that *the optimal function* $f_e(\rho) + \tilde{f}(\rho)$ *reaches its minimum at a unique point* ρ^0. The result is,

$$G(s) = \max_{0\le\rho^0\le 1} \left\{ e^{-s}\left[\rho^0 c(1-\rho^0)e^{-\frac{1}{2}\tilde{f}'(\rho^0)} + (\rho^0)^2(1-c)e^{\frac{1}{2}\tilde{f}'(\rho^0)}\right] - r(L\rho^0)/L \right\}, \tag{3.28}$$

where $\tilde{f}'(\rho^0)$ is determined by a condition

$$\tilde{f}'(\rho^0) + f_e{}'(\rho^0) = 0. \tag{3.29}$$

Finally, by using the explicit expression of $f_e(\rho)$ in it, we arrive at

$$G(s) = \max_{0\le\rho\le 1} \left\{ 2e^{-s}\sqrt{\rho^3(1-\rho)c(1-c)} - \left[\rho c(1-\rho) + (1-c)\rho^2\right] \right\}. \tag{3.30}$$

This type of variational principle, which describes the large-size behaviour of large deviation, is well known. As stated already, Garrahan et al. derived this formula and confirm the dynamical phase transition in [1–3]. Indeed, by drawing the variational function for several values of s, namely $s_1 < 0$, $s_2 = 0$, and $s_3 > 0$ in Fig. 3.1, we see that the argument ρ for realising the maximum value jumps at $s = 0$ suddenly. Due to this jump, the dynamical free energy $G(s)$ has a kink-like structure in the system size infinite limit as shown in the green line in Fig. 3.2. This first order phase transition is called the dynamical phase transition. Not only for this mean-field model, but also for several finite dimensional KCMs, this property was observed in Refs. [1–3] by using the population dynamics [13, 14], which is a numerical technique to obtain large deviation functions in time-series statistics.

For the interpretation of this property, we recall the biased ensemble introduced in Sect. 2.2.3. From the relation (2.14), which connects the derivative of $G(s)$ with the expected value of the activity in the biased ensemble, we can see that the singularity represents the biased ensemble to show the first order phase transition in the limit $L \to \infty$. Since, in this limit, the singular point is located at the origin $s = 0$, it indicates that the unbiased system ($s = 0$) has two expected values of the activity. One is 0, and the other one is (3.17), where the former represents inactiveness and latter does activeness. In other words, the unbiased system lies at the coexistence between the active and inactive phases.

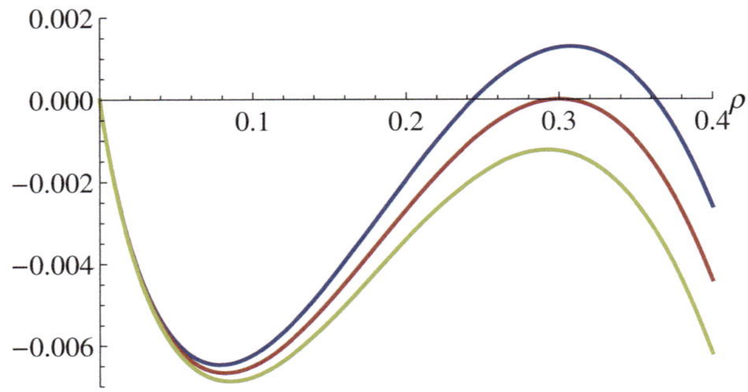

Fig. 3.1 The variational function in (3.30) as functions of ρ for several s. We set $c = 0.3$. The *blue*, *red*, and *yellow lines* correspond to $s = -0.01$, $s = 0$, and $s = 0.01$, respectively. One can see that ρ that maximises the function is located around $c = 0.3$ for $s < 0$, and located around 0 for $s > 0$. For $s = 0$, these two phases coexist, which means that there is a first order phase transition at $s = 0$

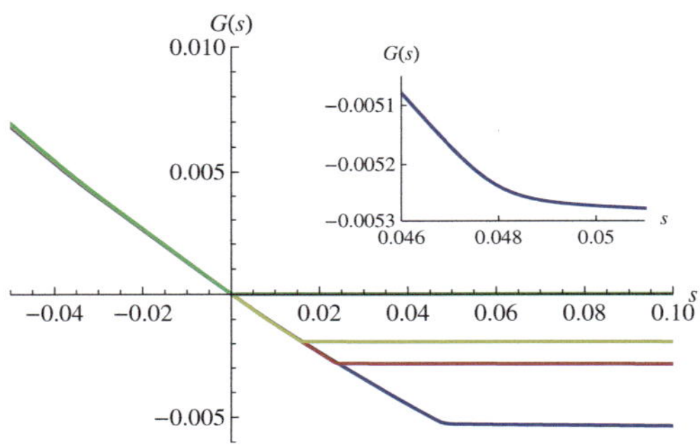

Fig. 3.2 Dynamical free energy $G(s)$ for $c = 0.3$ obtained from the variational formula (3.30) (*green*), and numerical diagonalisation of (3.20) for $L = 50$ (*blue*), $L = 100$ (*red*), $L = 150$ (*yellow*). The inset is a magnified picture of the *blue* line ($L = 50$) around the cusp, which shows the rounding of it due to the finite-size effect. We estimate the numerical examples of $s_c(L)$ defined by (3.31). For $L = 50$, $L = 100$, and $L = 150$, these are 0.0479... , 0.02390..., and 0.01591..., respectively. These values are close to $1/(2Lc(1-c))$ due to the formula (3.32). Indeed, $1/(2Lc(1-c))$ takes 0.04762..., 0.02381... and 0.01587... for $L = 50$, 100, and 150, respectively.

Finite Size Structure of the Dynamical Phase Transition

This first order phase transition gradually arises as the system size become larger. For seeing this, in Fig. 3.2, we show numerical examples of $G(s)$ for $c = 0.3$, $L = 50, 100, 150$ obtained by solving the largest eigenvalue problem (3.19). For each system size, we observe a cusp, even though the cross point is rounded as shown in the inset. This cross point becomes shaper as the system size becomes larger, and finally, in the system size infinite limit, $G(s)$ has an in continuity of the derivative at $s = 0$ (green line) as proved in the previous section.

For finite size system, even though the singularity is under developing, we can also define a remarkable point $s_c(L)$ by the special s that maximise the derivative of $G(s)$:

$$s_c(L) \equiv \underset{s}{\mathrm{Argmax}} G''(s). \tag{3.31}$$

As shown in Fig. 3.2, this point is not located at 0. For example, in this model, s_c scales like

$$s_c(L) = \frac{1}{(2c(1-c))} \frac{1}{L} + O\left(\frac{1}{L^2}\right), \tag{3.32}$$

where the derivation is in the end of Sect. 3.2.3. We show the numerical example in Fig. 3.2. In general KCMs, it is believed that $s_c(L)$ is scaled as the inverse of the system size [1–3]. The remarkable point is deviated from the origin. However, not like the equilibrium statistical physics, there are no direct correspondence of the biasing parameter s to physical field. (See Sect. 2.2). This has prevented the understanding of the dynamical phase transition. The purpose of this chapter is to shed a light on it with *the phenomenological structure for large deviation priniciple.* Indeed, with the formulation in Sect. 2.2.5, we can define the corresponding stationary state to the dynamical phase transition even if the system is finite, and the corresponding magnetization and susceptibility in it.

Free Energy, Modified Magnetization, and Modified Susceptibility

In the model we now consider, the unmodified (original) system satisfies the detailed balance condition (3.7). Furthermore, the quantity we consider is the dynamical activity defined as (3.14), which always takes 1. In this case, A detailed balance condition corresponding to the modified system directly follows from the definition of the modified transition rate $w^s(n \to n')$ (2.19) with (2.22). Indeed, $w^s(n \to n')$ satisfies

$$P^s(n) w^s(n \to n') = P^s(n') w^s(n' \to n) \tag{3.33}$$

with the equilibrium modified distribution $P^s(\boldsymbol{n})$ given as

$$P^s(n) = C P_{\mathrm{eq}}(n) \phi(n)^2, \tag{3.34}$$

where C is the normalisation constant and $\phi(n)$ is the left-eigenvector of (3.19). Here, from the distribution function $P^s(n)$, we define a modified free energy $F^s(n)$ and modifying free energy $\Delta F_s(n)$ as

$$F_s(n) \equiv -\log P_s(n), \tag{3.35}$$

and

$$\Delta F_s(n) \equiv -2 \log \phi(n). \tag{3.36}$$

There is a relationship between these two functions as

$$F_s(n) = -\log P_{\mathrm{eq}} + \Delta F_s(n) + \mathrm{const.} \tag{3.37}$$

Since we have the modified equilibrium distribution function, it might be also interesting to define a modified magnetization $\rho(s)$ and a modified susceptibility $\chi(s)$ as

$$\rho(s) = \sum_n (n/L) P^s(n), \tag{3.38}$$

$$\chi(s) = L \sum_n (n/L - \rho(s))^2 P^s(n). \tag{3.39}$$

In the following subsections, we approach to the dynamical phase transition by analyzing these modified properties.

3.2.2 Free Energy at $s = s_{\mathrm{c}}$ and the Finite-Size Correction

Numerical Examples of $\rho(s)$ and $\chi(s)$

We first show the numerical examples of $\rho(s)$ and $\chi(s)$ in Fig. 3.3. We solved the largest eigenvalue problem (3.19), then we constructed $P^s(n)$ from (3.34) and calculated $\rho(s)$ and $\chi(s)$. For the x-axis, we use a scaled variable $\tilde{s} = s/L$. In the figure, we can see the clear mark of the first order phase transition of the variable ρ around $\tilde{s} = 1/(2c(1-c))$, which becomes clearer as L becomes larger.

Numerical Examples of the Free Energies

To see the first order phase transition more clearly, we next see the free energies at the transition point. In Fig. 3.4, we show the equilibrium free energy $-\log P_{\mathrm{eq}}(n)/L$, the modifying free energy $\Delta F_s(n)/L$, and the modified free energy $F_s(n)/L$ for $s = 0.95 s_{\mathrm{c}}$, $s = s_{\mathrm{c}}$ and $s = 1.05\, s_{\mathrm{c}}$. The x-axis represents the density $\rho = n/L$. At $s = s_{\mathrm{c}}$, we observe that the modified free energy $F_s(n)$ reaches its minimum value at the two densities ρ_0 and ρ_1. One density characterises the inactiveness ($\rho \simeq 0$), and the other one does activeness ($\rho \simeq c$).

Analytical Expression of the Free Energy in $L \to \infty$ Limit

Now, we derive the analytical expression of the free energy, which might give us some insights behind the dynamical phase transition. We start with the eigenvalue equation (3.19) at the transition point $s = s_{\mathrm{c}}(L)$. In this expression, what we do is

1. assuming the large deviation property for the left-largest eigenvector $\phi(n)$:

$$\phi(n)|_{s=s_{\mathrm{c}}(L)} \sim e^{-L\Delta f_{s_{\mathrm{c}}(L)}(n/L)/2}, \tag{3.40}$$

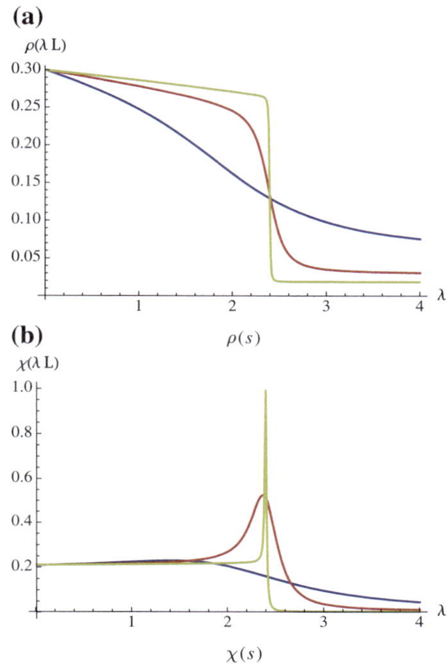

Fig. 3.3 The modified magnetisation $\rho(s)$ (**a**) and the modified susceptibility $\chi(s)$ (**b**) given in (3.38) and (3.39) as functions of $\lambda = s/L$. We numerically solve the largest eigenvalue problem (3.19), construct the distribution function $P^s(n)$ from (3.34), and calculate $\rho(s)$ and $\chi(s)$. We set $c = 0.3$, and $L = 20$ (*blue*), $L = 40$ (*red*), $L = 60$ (*yellow*). We note that the position of the peak of $\chi(s)$ is equal to $s_c(L)$ defined in (3.31), so that it close to $\lambda_c = 1/(2c(1-c))$ due to (3.32). Indeed, from the figure, we estimate it as $\lambda = 2.398...$ for $L = 60$, to which the corresponding λ_c is 2.381...

where $\Delta f_s(\rho)$ is given as

$$\Delta f_s(\rho) = \lim_{L\to\infty} \frac{1}{L} \Delta F_{s_c}(\rho L) \tag{3.41}$$

2. evaluating the leading order in $L \to \infty$ limit for the (left) largest eigenvalue problem (3.19).

Indeed, from $s = s_c(L)$, we can set

$$LG(s_c) = -c + O(1/L), \tag{3.42}$$

where we used (3.32). This leads to

$$\begin{aligned} &\phi(n+1)\tfrac{cn}{L}(L-n)\mathrm{e}^{-s_c} + \phi(n-1)\tfrac{(1-c)}{L}n(n-1)\mathrm{e}^{-s_c} \\ &-\phi(n)\left[c\tfrac{n}{L}(L-n) + (1-c)\tfrac{n}{L}(n-1) - c + O(1/L)\right] = 0. \end{aligned} \tag{3.43}$$

Then, from the assumption (3.40), we replace $\phi(n)$ in (3.43) by $\Delta f_{s_c}(\rho)$. The leading term in $L \to \infty$ limit is

$$\begin{aligned} &\mathrm{e}^{-(1/2)\partial\Delta f_{s_c}(\rho)/\partial\rho} c\rho(1-\rho) + \mathrm{e}^{(1/2)\partial\Delta f_{s_c}(\rho)/\partial\rho}(1-c)\rho^2 \\ &- \left[c\rho(1-\rho) + (1-c)\rho^2\right] = 0. \end{aligned} \tag{3.44}$$

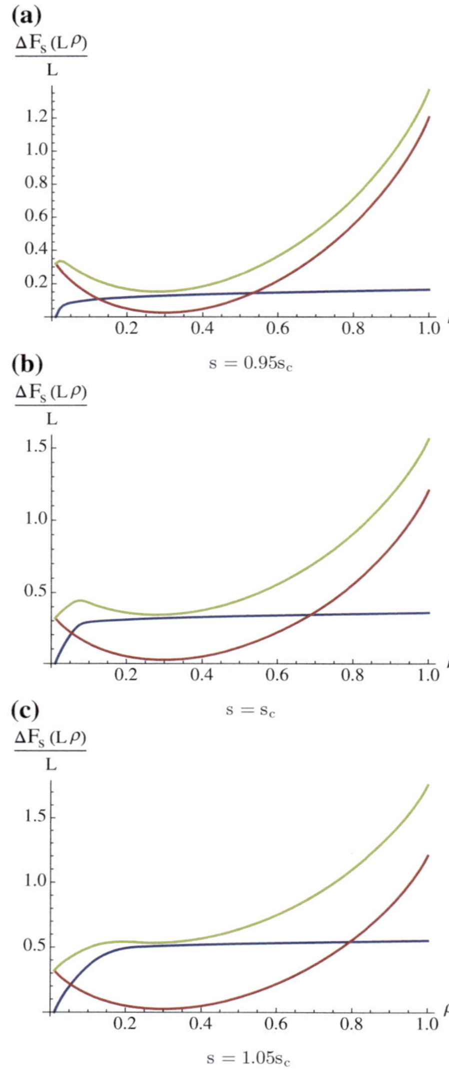

Fig. 3.4 Numerical examples of the modifying free energy $\Delta F_s(\rho L)/L$ (*blue*), the modified free energy $F_s(n)/L$ (*yellow*), and the original equilibrium free energy $-\log P_{\mathrm{eq}}(\rho L)/L$ (*red*) for $c = 0.3$, $L = 100$. We set $s = 0.95 s_{\mathrm{c}}$ (**a**), $s = s_{\mathrm{c}}$ (**b**) and $s = 1.05 s_{\mathrm{c}}$ (**c**). © IOP Publishing. Reproduced with permission. All rights reserved

This differential equation can be solved easily: There are two solutions,

$$\partial \Delta f_{s_{\mathrm{c}}}(\rho)/\partial \rho = 0 \tag{3.45}$$

and

$$\partial \Delta f_{s_{\mathrm{c}}}(\rho)/\partial \rho = -2 \log \left[\frac{(1-c)\rho}{c(1-\rho)} \right] = -2 \frac{\partial f_{\mathrm{e}}(\rho)}{\partial \rho}, \tag{3.46}$$

which are equivalent to

$$\Delta f_{s_c}(\rho) = \text{const.}, \tag{3.47}$$

and

$$\Delta f_{s_c}(\rho) = -2f_e(\rho) + \text{const.} \tag{3.48}$$

Next, by looking at the numerical example of ΔF_{s_c} in Fig. 3.4 (the blue line), we expect that these two solutions are connected non-analytically. That is, with a connecting point ρ_c^∞, $\Delta f_{s_c}(\rho)$ is given as

$$\Delta f_{s_c}(\rho) = -2f_e(\rho) + 2f_e(0). \tag{3.49}$$

for $\rho \le \rho_c^\infty$ and

$$\Delta f_{s_c}(\rho) = -2f_e(\rho_c^\infty) + 2f_e(0). \tag{3.50}$$

for $\rho > \rho_c^\infty$. The connecting point ρ_c^∞ is determined from the condition of first order phase transition, which means that the inactive ($\rho \simeq 0$) and the active ($\rho \simeq c$) configurations have the same weight:

$$f_e(c) + \Delta f_{s_c}(c) = f_e(0) + \Delta f_{s_c}(0). \tag{3.51}$$

With the explicit expression of f_e and Δf_{s_c}, we obtain an equation for ρ_c^∞ as

$$2\left[(1-\rho_c^\infty)\log\frac{1-\rho_c^\infty}{1-c} + \rho_c^\infty \log\frac{\rho_c^\infty}{c}\right] = -\log(1-c). \tag{3.52}$$

We show the solution of this equation as a function of c in Fig. 3.5. Then, in Fig. 3.6, we plot the obtained $\Delta f_{s_c}(\rho)$ for $c = 0.3$. On the same figure, we also plot $f_e(\rho)$ and $f_{s_c}(\rho)$. We find that the distribution of the density ρ is naturally divided into two domains by ρ_c^∞. The active domain given as $\rho > \rho_c^\infty$ has the same distribution as the unbiased system, whears the inactive domain given as $\rho < \rho_c^\infty$ has a deep trap at the origin $\rho \sim 0$.

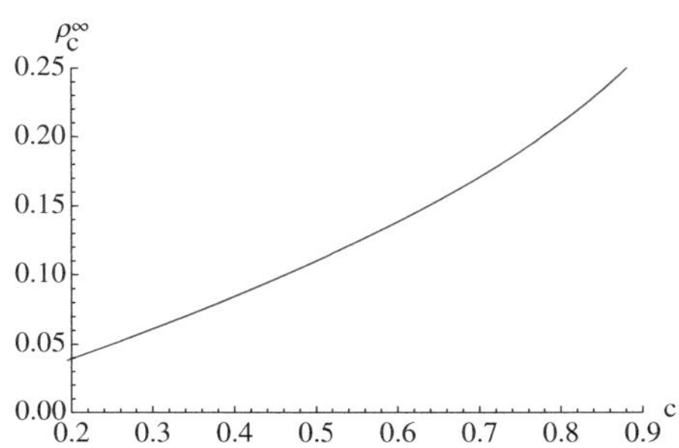

Fig. 3.5 Connecting point ρ_c^∞ given as (3.52). We numerically solve (3.52) and plot it as a function of c. © IOP Publishing. Reproduced with permission. All rights reserved

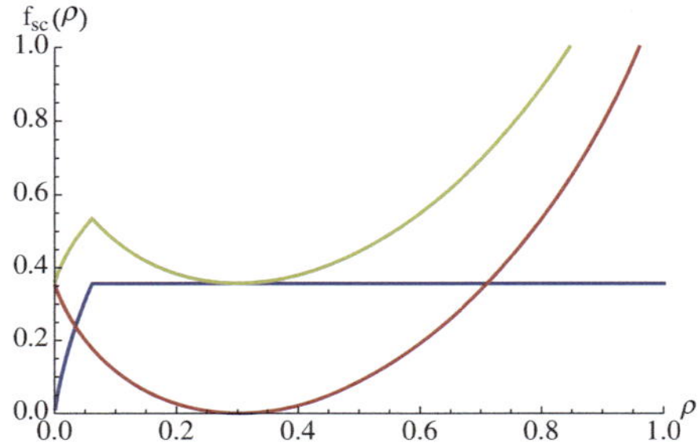

Fig. 3.6 Analytical expression of the modifying free energy for $s = s_c$ given as (3.49) and (3.50). We plot it with *blue line*. We also plot the corresponding modified free energy given as $f_e(\rho) + \Delta f_{s_c}(\rho)$ with *yellow line*, and the unbiased equilibrium free energy $f_e(\rho)$ with *red line*.

Finite Size Correction

It is hard to diagonalise the matrix (3.19) for sufficiently large L to check the obtained expression of $\Delta f_{s_c}(\rho)$. Here, we seek for the finite size correction of $\Delta f_{s_c}(\rho)$ in order to confirm our analytical expression. It will be turned out that those corrections will be used to derive the finite-size scaling of the large deviation function, as explained in the next subsection.

First, for finite-size systems, we define a finite-size connecting point ρ_c^L as

$$\sum_{n \le n_c^L} P^s(n) = \sum_{n > n_c^L} P^s(n), \tag{3.53}$$

where $n_c = \lfloor L\rho_c^L \rfloor$. Here, we note that the Eq. (3.53) can give a non-integer value of ρ_c^L, because the distribution function $P^s(n)$ has a ρ_c^L dependence as shown below. To obtain the next order correction of $\Delta f_{s_c}(\rho)$, we use a perturbation analysis. More precisely, for each active and inactive region, we assume the following scaling form

$$\Phi_L(n) = \exp\left\{-(L/2)\left[\Delta f_{s_c}(n/L) + (1/L)\Delta f_{s_c}^{(1)}(n/L)\right]\right\}, \tag{3.54}$$

substitute it in the eigenvalue equation (3.43), and derive the equation for $\Delta f_{s_c}^{(1)}(n/L)$. We don't show the precise derivation here. That is done in Appendix B.1. The result is

$$\Delta f_{s_c}^{(1)}(\rho) = -\log\frac{\rho(1-\rho)}{(c-r)^2} - \frac{\rho(1-2c)}{c(1-c)} + \text{const.} \tag{3.55}$$

for $\rho \le \rho_c^L$, and

$$\Delta f_{s_c}^{(1)}(\rho) = -2\left[\frac{\rho(2c-1)}{2c(1-c)} - \log\rho\right] + \text{const.} \tag{3.56}$$

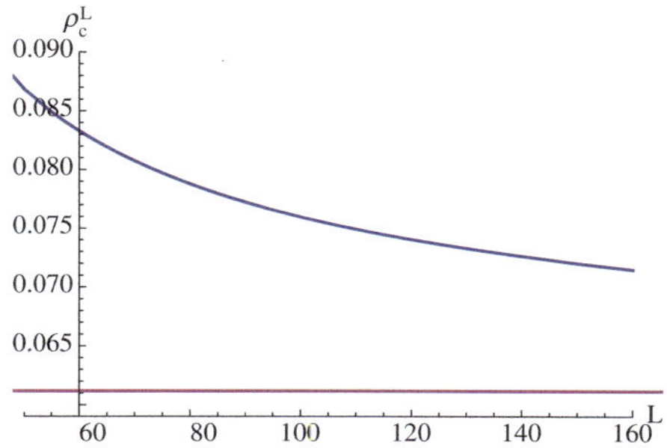

Fig. 3.7 Numerical example of ρ_c^L (*blue*) obtained by solving (3.53) as a function of L with its infinite limit $\rho_c^\infty = \lim_{L\to\infty} \rho_c^L$ (*purple*) determined by (3.52). We set $c = 0.3$. The figure shows the slow convergence of ρ_c^L to ρ_c^∞.

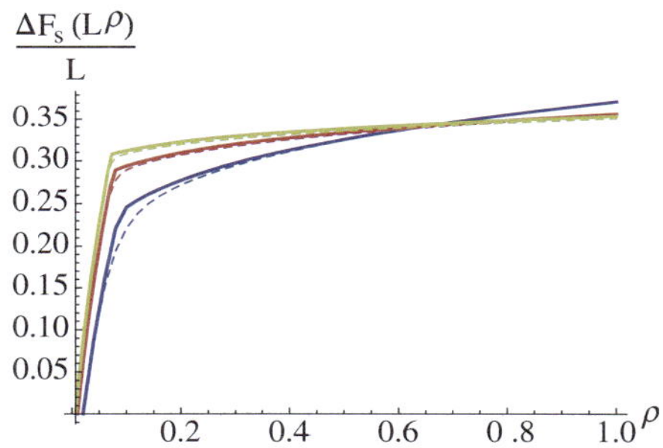

Fig. 3.8 Numerical check of the analytical expression of the modifying free energy for finite-size system given as $\Delta f_{s_c}(n/L) + (1/L)\Delta f_{s_c}^{(1)}(n/L)$. We set $c = 0.3$, $s = s_c$, $L = 50$ (*blue*), $L = 100$ (*red*) and $L = 150$ (*yellow*). We plot analytical expression $\Delta f_{s_c}(n/L) + (1/L)\Delta f_{s_c}^{(1)}(n/L)$ in *solid lines*, and also $\Delta F_s(L\rho)/L$ obtained from numerical diagonalisation of $L^s_{n,n'}$ with *dashed lines*.

for $\rho > \rho_c^L$, where the two constants are determined by the conditions $\Delta f_{s_c}^{(1)}(1/L) = 0$ and $\lim_{\rho\to\rho_c^L} \Delta f_{s_c}^{(1)}(\rho) = \lim_{\rho\to\rho_c^L} \Delta f_{s_c}^{(1)}(\rho)$. We note that the latter constant depends on ρ_c^L, so that $P^s(n)$ $(n > n^L)$ does as well. This ensures that ρ_c^L defined in (3.53) is not a trivial-integer as shown in Fig. 3.7, where we plot the numerical examples of ρ_c^L as a function of L together with ρ_c^∞.

Finally, in Fig. 3.8, we plot our finite-size free energy with the numerical results obtained by direct diagonalization of $L^s_{n,n'}$ for $c = 0.3$, $s = s_c$, $L = 50, 100, 150$. We can see a clear agreement between them as L increases.

3.2.3 Scaling Function Around $s = s_c$

In the previous subsection, we observed that the system showed a phase coexistence for the density ρ, which was similar to the well-known equilibrium phase transition although the mechanism behind it might be different. In this subsection, for more deep understanding of the dynamical phase transition, apart from the original motivation about the glassiness of the system, we study more precisely what is the difference between the dynamical phase transition and equilibrium phase transition. Especially, we focus on the scaling function around the transition point. For the equilibrium ferromagnetic model, the scaling functions around the 1st order phase transition point have been determined by Borgs and Kotecký in [8, 9]. In the paper, they derived a scaling function by assuming that the distribution function at the phase coexistence point is written as the sum of the distribution function of each phase. That scaling function was universal for general ferromagnet model. What we consider below is if this universality is valid or not even for the case of dynamical phase transition. Then, if it is not valid, the next question would be how we modify the method to apply to those dynamical phase transitions.

Numerical Confirmation of the Existence of Scaling Functions

We first numerically confirm the existence of the scaling functions. For this, we define the *width* of the phase transition region for finite size systems. As seen in Fig. 3.3, for a given L, the width of the phase transition region is inversely proportional to the derivative of $\rho(s)$ at s_c. We thus define a scaling ratio κ as

$$\kappa = -\left.\frac{\partial \rho(s)}{\partial s}\right|_{s=s_c}. \tag{3.57}$$

We plot the logarithm of κ for various values of c as a function of L in Fig. 3.9. From the figure, we find that $\log \kappa$ is proportional to L. This means that the width of the first order coexistence region shrinks with an exponentially fast speed as the system size L becomes larger.

Next, by using κ, we define a scaled variable x as

$$x = \kappa(s - s_c(L)). \tag{3.58}$$

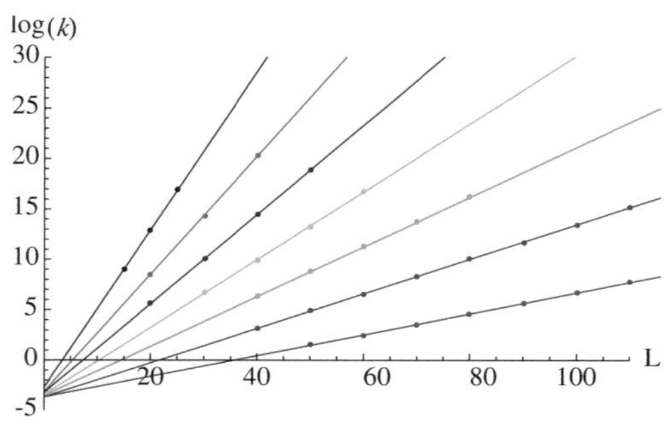

Fig. 3.9 The logarithm of the scaling ratio κ defined in (3.57) as a function of L for $c = 0.2$ (*red*), $c = 0.3$ (*blue*), $c = 0.4$ (*green*), $c = 0.5$ (*yellow*), $c = 0.6$ (*purple*), $c = 0.7$ (*gray*) and $c = 0.8$ (*black*). We solved the eigenvalue equation of $L^s_{n,n'}$ and calculate κ for several L and c. These are plotted as *dots*. *Solid lines* are the liner fit obtained from those data points.

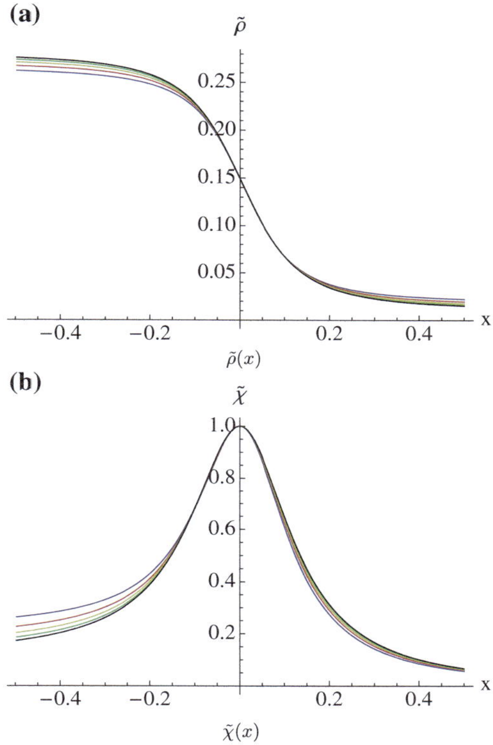

Fig. 3.10 Numerical examples of the rescaled magnetisation $\tilde{\rho}(x)$ (**a**) and the rescaled susceptibility $\tilde{\chi}(x)$ (**b**). We set $c = 0.3$, and $L = 60$ (*blue*), $L = 70$ (*red*), $L = 80$ (*green*), $L = 90$ (*yellow*), and $L = 100$ (*black*).

Also, we define rescaled magnetisation and susceptibility as

$$\tilde{\rho}(x) = \rho\left(x\kappa^{-1} + s_c(L)\right), \tag{3.59}$$

and

$$\tilde{\chi}(x) = \frac{\chi\left(x\kappa^{-1} + s_c(L)\right)}{\chi(s_c(L))}. \tag{3.60}$$

These definitions represent the magnetisation and susceptibility magnified around the dynamical phase transition point. We plot these functions in Fig. 3.10 for large values of L. Converging of the resealed function is observed in the figure.

Ansatz to Determine the Analytical Expression of the Scaling Properties

Here, we propose an ansatz to determine the analytical expression of the scaling properties. Before showing it, we first recall the finite-size free energy at the coexistence point $s = s_c$. With $\Delta f_{s_c}(\rho)$ and $\frac{1}{L}\Delta f_{s_c}^{(1)}(\rho)$ given as (3.49), (3.50), (3.55), and (3.56), we obtained the active and inactive finite-size modifying free energy f_i and f_a as

$$f_i(\rho) = \Delta f_{s_c}(\rho) + \frac{1}{L}\Delta f_{s_c}^{(1)}(\rho) \tag{3.61}$$

for $\rho \leq \rho_c^L$

$$f_a(\rho) = \Delta f_{s_c}(\rho) + \frac{1}{L}\Delta f_{s_c}^{(1)}(\rho) \tag{3.62}$$

for $\rho > \rho_c^L$. Then, the corresponding distribution function is

$$P^{s_c}(n) = P_i(n)\,\mathbf{1}_{n\leq n_c^L} + P_a(n)\,\mathbf{1}_{n>n_c^L}, \tag{3.63}$$

where $P_i(n)$ and $P_a(n)$ are defined as

$$P_i(n) = \frac{1}{Z_i + Z_a} P_{eq}(n)e^{-Lf_i(n/L)} \tag{3.64}$$

$$P_a(n) = \frac{1}{Z_i + Z_a} P_{eq}(n)e^{-Lf_a(n/L)} \tag{3.65}$$

with the normalization constant

$$Z_i = \sum_{n\leq n_c} P_{eq}(n)e^{-Lf_i(n/L)}, \tag{3.66}$$

$$Z_a = \sum_{n>n_c} P_{eq}(n)e^{-Lf_a(n/L)}. \tag{3.67}$$

By using the distribution function at $s = s_c$, we approach to the problem around s_c ($s \sim s_c$). Our ansatz to determine the scaling function is composed of following two parts:

1. For around s_c ($s \sim s_c$), we assume that the distribution function is given as

$$P^s(n) = [1 + a^*(s)]\delta_{n\leq n_c} P_i(n) + [1 - a^*(s)]\delta_{n>n_c} P_a(n), \tag{3.68}$$

   where $a^*(s)$ is a 'mixing function' to determine. We note that the distribution function satisfies the normalization condition due to $Z_i = Z_a$.
2. The mixing function $a^*(s)$ is determined from the variational principle (2.29) for the corresponding stationary state to biased ensemble.

For the first part, this definition of the distribution function is inspired by the method used by Borgs and Kotecký for obtaining the scaling function in equilibrium first order phase transition of ferromagnet [8, 9]. However, we should mention that our distribution function $P^s(n)$ doesn't describe a superposition between two elementary distributions not like the method by them. Rather, our method represents a separation of the domain of occupation number n into two phases. The second part is specific to our method, because there are no corresponding variational principle in equilibrium statistical mechanics. For equilibrium statistical mechanics, the biased ensemble directly corresponds to another equilibrium distribution function, whereas for time-series statistics, we need to use the variational principle (2.29) to create the corresponding another distribution function as explained in Sect. 2.2.5.

Here we recall the variational formula for obtaining $G(s)$, (3.30), derived in Sect. 3.2.1. Naively, one might think a blunt perturbation in $1/L$ around this formula is working. However, it isn't true. This is mainly due to the reason that there is a 1st order phase transition so that the saddle point is not only one but two. Our method contains this fact as the first part of the assumption. In other words, by combining the finite size scaling method by Borgs and Kotecký with the variational principle, we overcome this difficulty. To the best of our knowledge, this ansatz hasn't been used up to now.

Determination of $a^*(s)$

Here, we show the details of the calculation for determining $a^*(s)$. First, from the ansatz (3.68) with the relation (3.34), we construct $\phi(n)$ as

$$\phi(n) \propto \delta_{n\le n_c}\sqrt{1+a^*(s)}\mathrm{e}^{-Lf_\mathrm{i}(n/L)/2} + \delta_{n>n_c}\sqrt{1-a^*(s)}\mathrm{e}^{-Lf_\mathrm{a}(n/L)/2}, \tag{3.69}$$

also the magnetisation $\rho(s)$, susceptibility $\chi(s)$ around s_c as

$$\rho(s) = \frac{\langle\rho\rangle_\mathrm{i}}{2}\left[1+a^*(s)\right] + \frac{\langle\rho\rangle_\mathrm{a}}{2}\left[1-a^*(s)\right], \tag{3.70}$$

$$\chi(s) = L\left\{\frac{\langle\rho^2\rangle_\mathrm{i}}{2}\left[1+a^*(s)\right] + \frac{\langle\rho^2\rangle_\mathrm{a}}{2}\left[1-a^*(s)\right] - \rho(s)^2\right\}, \tag{3.71}$$

where $\langle\ \rangle_\mathrm{i}$ and $\langle\ \rangle_\mathrm{a}$ are the expected values in the active and the inactive phases defined as

$$\langle g\rangle_\mathrm{i} = 2\sum_{n\le n_c} P_\mathrm{i}(n)g(n) \tag{3.72}$$

and

$$\langle g\rangle_\mathrm{a} = 2\sum_{n>n_c} P_\mathrm{a}(n)g(n), \tag{3.73}$$

respectively. These formulas expresses that we can calculate every scaling property around s_c from $a^*(s)$.

In order to determine $a^*(s)$, we use the variational principle (2.29). By combining it with the transition rate $w(n\to n')$, the stationary distribution function (3.68), and $\phi(n)$ given as (3.69), we obtain the following equations:

$$\partial\Psi(a)/\partial a|_{a=a^*(s)} = 0 \tag{3.74}$$

with

$$\Psi(a) = \frac{1}{L}\sum_n P^s(n)\left[\tilde{r}(n) - r(n)\right], \tag{3.75}$$

where $\tilde{r}(n)$ is defined as

$$\tilde{r}(n) = nc\left(1 - \frac{n}{L}\right)\frac{\phi(n+1)}{\phi_{\mathrm{L}}(n)}\mathrm{e}^{-s} + n(1-c)\left(\frac{n}{L} - \frac{1}{L}\right)\frac{\phi(n-1)}{\phi(n)}\mathrm{e}^{-s}. \quad (3.76)$$

We show the details of the calculation in Appendix B.2. Here, we just write down the result. The answer of (3.74) is

$$a^*(s) = \frac{A}{\sqrt{1+A^2}}, \quad (3.77)$$

$$A = (\Omega_< - \Omega_>)/\Omega_=, \quad (3.78)$$

where $\Omega_<$, $\Omega_>$, and $\Omega_=$ are defined as

$$\Omega_= = 2\frac{n_{\mathrm{c}}}{L}c(1 - \frac{n_{\mathrm{c}}}{L})P_{\mathrm{i}}(n_{\mathrm{c}})\frac{\mathrm{e}^{-Lf_{\mathrm{a}}((n_{\mathrm{c}}+1)/L)/2}}{\mathrm{e}^{-Lf_{\mathrm{i}}(n_{\mathrm{c}}/L)/2}}\mathrm{e}^{-s}, \quad (3.79)$$

$$\Omega_< = \frac{1}{2L}\left\langle \tilde{r}_{\mathrm{i}}\mathrm{e}^{-s} - r\right\rangle_{\mathrm{i}}, \quad (3.80)$$

$$\Omega_> = \frac{1}{2L}\left\langle \tilde{r}_{\mathrm{a}}\mathrm{e}^{-s} - r\right\rangle_{\mathrm{a}}, \quad (3.81)$$

with the definition of $\tilde{r}_{\mathrm{i,a}}(n)$ as

$$\tilde{r}_{\mathrm{i,a}}(n) = nc\left(1 - \frac{n}{L}\right)\frac{\mathrm{e}^{-Lf_{\mathrm{i,a}}((n+1)/L)/2}}{\mathrm{e}^{-Lf_{\mathrm{i,a}}(n/L)/2}} + n(1-c)\left(\frac{n}{L} - \frac{1}{L}\right)\frac{\mathrm{e}^{-Lf_{\mathrm{i,a}}((n-1)/L)/2}}{\mathrm{e}^{-Lf_{\mathrm{i,a}}(n/L)/2}}.$$

Analytical Expression of (1/L) log κ

From now on, by using the obtained $a^*(s)$, we discuss the scaling properties. We first obtain the analytical expression of $(1/L)\log\kappa$, where the κ is the scaling ratio define by (3.57).

From $a^*(s_{\mathrm{c}}) = 0$, we have

$$\Omega_<|_{s=s_{\mathrm{c}}} = \Omega_>|_{s=s_{\mathrm{c}}}. \quad (3.82)$$

Then, by expanding this equation around $s = s_{\mathrm{c}}$, we obtain

$$A = -\frac{s - s_{\mathrm{c}}}{\Omega_=^{\mathrm{c}}}\left[\left\langle\frac{r}{2L}\right\rangle_{\mathrm{i}} - \left\langle\frac{r}{2L}\right\rangle_{\mathrm{a}}\right] + O((s-s_{\mathrm{c}})^2), \quad (3.83)$$

where we denote $\Omega_=|_{s=s_{\mathrm{c}}}$ by $\Omega_=^{\mathrm{c}}$. From the definition of $\kappa = -\partial\rho(s)/\partial s|_{s=s_{\mathrm{c}}}$ and (3.70), we thus find the analytical expression of it as

$$\kappa = \frac{1}{\Omega_{=}^{c}} \left[\left\langle \frac{\rho}{2} \right\rangle_{i} - \left\langle \frac{\rho}{2} \right\rangle_{a} \right] \left[\left\langle \frac{r}{2L} \right\rangle_{i} - \left\langle \frac{r}{2L} \right\rangle_{a} \right]. \tag{3.84}$$

To see the exponential dependence of κ, we recall that $P_{i}(n_{c})$ in $\Omega_{=}^{c}$ goes to 0 in $L \to \infty$ with exponentially fast speed due to the large deviation principle, wheras the other terms converges to each corresponding value in the limit. By combining it with the explicit expression of the free energy $f_{i}(\rho)$, we thus arrive at

$$\lim_{L\to\infty} \frac{1}{L} \log\kappa = f_{i}(\rho_{c}^{\infty}) = -\frac{1}{2} \log(1-c). \tag{3.85}$$

In the Fig. 3.11, we plot the right-hand side of (3.85), $-\frac{1}{2}\log(1-c)$, as a function of c (blue line). On the same figure, we plot the slope of the straight line in Fig. 3.11 for $c = 0.2, 0.3, \ldots, 0.8$ (red dots), which corresponds to the left-hand side of (3.85). We can see the coincide very well.

Our formula claims that $(1/L)\log\kappa$ converges to the height of the large deviation function from the bottom to the connecting point ($\rho = \rho_{c}^{\infty}$). We note that this fact reminds us the instantonic approach in [10] by Jörg, Krzakala, Kurchan, Maggs, Pujos, and in [11] by Bapst, Semerjian for obtaining the exponentially small gap in the quantum ferromagnet. In the next section, by bridging the time-series statistics to quantum systems, we re-derive their formula by using our approach.

Analytical Expressions of $\tilde{\rho}(x)$ and $\tilde{\chi}(x)$

Next, we obtain $\tilde{\rho}(x)$ and $\tilde{\chi}(x)$. From the definition of x in (3.58) with (3.83) and (3.84), we have

$$A = \frac{2x}{\langle\rho\rangle_{a} - \langle\rho\rangle_{i}} + O(\kappa^{-1}). \tag{3.86}$$

Then, by combining it with (3.70), (3.71) and (3.77), we obtain

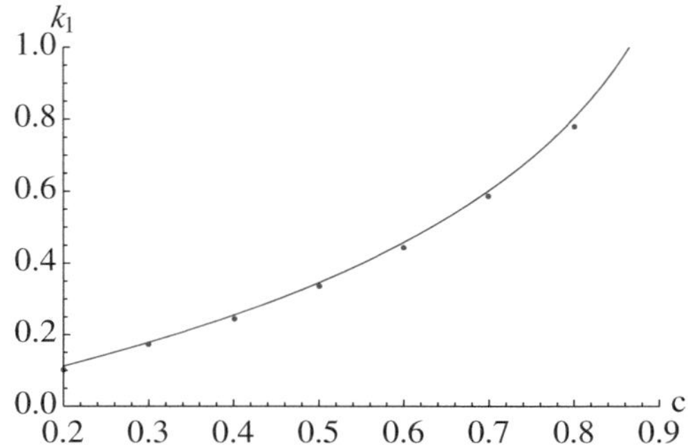

Fig. 3.11 Numerical check of the formula giving κ (3.85). We estimated $(1/L)\log\kappa$ for $c = 0.2, 0.3, \ldots, 0.8$ from the fitting lines in Fig. 3.9 and plot them with *red dots*. On the same figure, we draw the analytical result $-\frac{1}{2}\log(1-c)$ with *blue line*. The agreement is relatively good although *red dots* are obtained from finite size L.

$$\tilde{\rho}(x) = \frac{1}{2}\left[\langle\rho\rangle_{\mathrm{i}} + \langle\rho\rangle_{\mathrm{a}} - \frac{2x}{\sqrt{1+4x^2\left[\langle\rho\rangle_{\mathrm{i}} - \langle\rho\rangle_{\mathrm{a}}\right]^{-2}}}\right], \tag{3.87}$$

$$\begin{aligned}\tilde{\chi}(x) &= \frac{1}{\left\langle\rho^2\right\rangle_{\mathrm{i}} + \left\langle\rho^2\right\rangle_{\mathrm{a}} - \left[\langle\rho\rangle_{\mathrm{i}} + \langle\rho\rangle_{\mathrm{a}}\right]^2/2} \\ &\times\left[\left\langle\rho^2\right\rangle_{\mathrm{i}} + \left\langle\rho^2\right\rangle_{\mathrm{a}} - \frac{2x\left[\left\langle\rho^2\right\rangle_{\mathrm{i}} - \left\langle\rho^2\right\rangle_{\mathrm{a}}\right]\left[\langle\rho\rangle_{\mathrm{i}} - \langle\rho\rangle_{\mathrm{a}}\right]^{-1}}{\sqrt{1+4x^2\left[\langle\rho\rangle_{\mathrm{i}} - \langle\rho\rangle_{\mathrm{a}}\right]^{-2}}} - 2\tilde{\rho}(x)^2\right],\end{aligned} \tag{3.88}$$

where we omit the exponentially small term $O(\kappa^{-1})$. Finally, by noticing $\lim_{L\to\infty}\langle\rho\rangle_{\mathrm{i}} = 0$ and $\lim_{L\to\infty}\langle\rho\rangle_{\mathrm{a}} = c$ from the free energies (3.49) and (3.50), we arrive at the scaling functions as

$$\tilde{\rho}_{\infty}(x) = \lim_{L\to\infty}\tilde{\rho}(x) = \frac{1}{2}\left[c - \frac{2x}{\sqrt{1+4x^2c^{-2}}}\right], \tag{3.89}$$

$$\tilde{\chi}_{\infty}(x) = \lim_{L\to\infty}\tilde{\chi}(x) = \frac{c^2}{c^2+4x^2}. \tag{3.90}$$

For the numerical conformation, we use the expressions (3.87) and (3.88), which contains the finite-size correction of (3.89) and (3.90). These expressions are also constituted of the expected value and the variance of ρ in each of the active and inactive phases, so that it is suggestive about its universal form. We plot (3.87) and (3.88) (red solid lines) in Fig. 3.12 with the corresponding numerical results (blue dashed lines). We can see the coincidence between them very well. On the same figure, we also plot the infinite size scaling functions (3.89) and (3.90) as yellow lines. Since there is a deviation between the yellow lines and the others, we understand the importance of the finite size corrections in the formulas (3.87) and (3.88).

Until now, we have focused on the scaling property of $\rho(s)$ and $\chi(s)$. As the corollary of the result, one can obtain the scaling functions for the expected value and the susceptibility of the activity given as $\partial G(s)/\partial s$ and $\partial^2 G(s)/\partial s^2$. It is worth mentioning that even though $\partial^2 G(s)/\partial s^2$ is not connected directly to the equilibrium distribution function P_s, an expression for the scaling functions can be derived thanks to the results (3.87) and (3.88). See Appendix B.3 for the details.

Derivation of the Analytical Expression of s_c

By using our formulation, we also can derive an analytical expression of s_{c} as

$$s_{\mathrm{c}} = \frac{1}{2Lc(1-c)} + O(1/L^2), \tag{3.91}$$

which is numerically confirmed in Fig. 3.2. Here, we show the derivation.

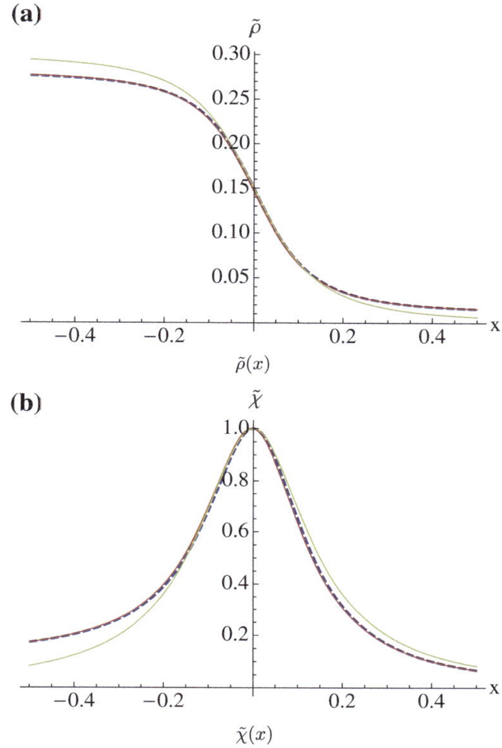

Fig. 3.12 Numerical check of the analytical expressions of the rescaled magnetisation $\tilde{\rho}(x)$ given in (3.87) (**a**) and the susceptibility $\tilde{\chi}(x)$ given in (3.88) (**b**). We set $c = 0.3$ and $L = 100$. The *solid red lines* are the analytical results, ant the *dashed blue lines* are the corresponding numerical results obtained from the diagonalisation of the matrix $L^s_{n,n'}$. On the same figure, for a comparison, we also plot the infinite-size scaling functions (3.89) and (3.90) with *yellow solid lines*. The deviation of *yellow lines* from the others show the importance of the finite size corrections in (3.87) and (3.88). © IOP Publishing. Reproduced with permission. All rights reserved

We evaluate the both sides of (3.82) up to $O(1/L)$ with the explicit results of r, $\tilde{r}_{\mathrm{i}}$, and $\tilde{r}_{\mathrm{a}}$. We first consider the left-hand side of (3.82). By taking the saddle point of the summation, we rewrite it as

$$\frac{1}{L}\left(\tilde{r}_{\mathrm{i}}\mathrm{e}^{-s_{\mathrm{c}}} - r\right)\Big|_{n=1} + O(1/L^2) = \frac{c}{L}\mathrm{e}^{-s_{\mathrm{c}}}\frac{\mathrm{e}^{-Lf_{\mathrm{i}}(2/L)/2}}{\mathrm{e}^{-Lf_{\mathrm{i}}(1/L)/2}} - \frac{c}{L} + O(1/L^2). \quad (3.92)$$

Here, the first term is $O(1/L^2)$ because

$$\frac{\mathrm{e}^{-Lf_{\mathrm{i}}(2/L)/2}}{\mathrm{e}^{-Lf_{\mathrm{i}}(1/L)/2}} = O(1/L). \quad (3.93)$$

We thus find that the left-hand side of (3.82) is

$$-c/L + O(1/L^2). \quad (3.94)$$

Then, on the other side, the right-hand side of (3.82) becomes

$$\begin{aligned}&\tfrac{1}{L}\left(\tilde{r}_{\mathrm{a}}\mathrm{e}^{-s_{\mathrm{c}}}-r\right)\Big|_{n=Lc}+O(1/L^2)\\ &=c^2(1-c)\left[\mathrm{e}^{-\frac{1}{2}\frac{\partial fa(\rho)}{\partial\rho}|_{\rho=c}-s_{\mathrm{c}}}+\mathrm{e}^{\frac{1}{2}\frac{\partial fa(\rho)}{\partial\rho}|_{\rho=c}-s_{\mathrm{c}}}-2\right]+O(1/L^2).\end{aligned} \tag{3.95}$$

due to the saddle point approximation. Here, we notice that $\partial f_a(\rho)/\partial\rho|_{\rho=c}$ and s_{c} are $O(1/L)$. We thus we rewrite (3.95) as

$$-2s_{\mathrm{c}}c^2(1-c)+O(1/L^2). \tag{3.96}$$

Therefore, by comparing the both sides of (3.82), that is, (3.94) and (3.96), we finally arrive at

$$s_{\mathrm{c}}=\frac{1}{2Lc(1-c)}+O(1/L^2). \tag{3.97}$$

3.3 Mean-Field Quantum Ferromagnet and the Scaling Function

For time-series statistics, when the system satisfies the detailed balance condition, the largest eigenvalue problem (3.20) can be symmetrized. This indicates that the mathematical structure behind the largest eigenvalue problem for time-series statistics, and the eigenvalue of problem of the ground state are common. For example, the variational principle (3.21) and the variational principle for the ground state in quantum systems are equivalent as shown in Appendix B.4. With keeping this correspondence in our mind, here in this section, we apply our finite-size scaling method to a mean-field quantum ferromagnet. Especially, we focus on the quantum phase transition in the model. As a result, we find the same scaling functions as the one for dynamical phase transition. Furthermore, to this system, by applying our formula (3.85), which was used for obtaining the width of the coexistence region of the dynamical phase transition in the previous section, we obtain the same expression of the formula that gives the exponentially small energy gap between the ground state and the first excited state. This is exactly the same as the one that was obtained by Bapst and Semerjian in Ref. [11].

3.3.1 *Preliminary*

Definition of Model

We consider L interacting 1/2 spins, where the Hilbert space is spanned by the space $\{|\vec{\sigma}\rangle \mid \vec{\sigma}=(\sigma_1,\ldots,\sigma_L)\in\{-1,+1\}^L\}$. The Pauli matrices acting on the ith spin are denoted by $\hat{\sigma}_i^x$, $\hat{\sigma}_i^y$, and $\hat{\sigma}_i^z$. They satisfy $\sigma_i^z|\vec{\sigma}\rangle=\sigma_i|\vec{\sigma}\rangle$, $\sigma_i^x|\vec{\sigma}\rangle=|\vec{\sigma}^{(i)}\rangle$, where $\vec{\sigma}^{(i)}$ is the configuration in which the i-th spin is flipped. We define the transverse magnetisation and the longitudinal magnetisation as

$$\hat{m}^x = \frac{1}{L}\sum_{i=1}^{L}\hat{\sigma}_i^x, \tag{3.98}$$

$$\hat{m}^z = \frac{1}{L}\sum_{i=1}^{L}\hat{\sigma}_i^z. \tag{3.99}$$

By using these definitions, the Hamiltonian is defined as

$$\hat{H} = -L(\hat{m}^z)^p - \Gamma L\hat{m}^x. \tag{3.100}$$

This model is called a mean-field quantum p-spin ferromagnet. It is known that a quantum phase transition takes place in this model for a special value of Γ . We note that for the $p = 2$ (quantum Curie-Weiss model), the transition is second-order, whereas for the $p \geq 3$ the transition is first-order. See Ref. [11] for the details of the thermodynamic properties of this model. Hereafter, we focus on the case that $p \geq 3$.

Mapping from the Ground State Eigenvalue Problem to the Formulation in Time-Series Statistics

Here, we show a mapping of the ground state eigenvalue problem to the formulation in time-series statistics. Thanks to it, we will be able to apply the method developed in the previous method to the quantum phase transition in this system.

We denote the ground state of $\hat{H}$ by $|\Phi\rangle$ and the ground energy by E:

$$\hat{H}|\Phi\rangle = E|\Phi\rangle \tag{3.101}$$

Then, it is known that the ground state of $\hat{H}$ lies in a symmetries subspace, where the interchanges of two spins are permitted:

$$\langle\vec{\sigma}|\Phi\rangle = \Phi(m^z(\vec{\sigma})), \tag{3.102}$$

with a definition of $m^z(\vec{\sigma})$ as $m^z = (1/L)\sum_{i=1}^{L}\sigma_i^z$. See Ref. [11] for the example of the proof. By multiplying (3.101) by $\langle\vec{\sigma}|$ from the left, we obtain

$$-L(m^z)^p\langle\vec{\sigma}|\Phi\rangle - \Gamma L\frac{1}{L}\sum_{i=1}^{L}\langle\vec{\sigma}^{(i)}|\Phi\rangle = E\langle\vec{\sigma}|\Phi\rangle. \tag{3.103}$$

Then, by using (3.102), we rewrite it as

$$-L(m^z)^p\Phi(m^z) - \Gamma L\tfrac{1}{L}\textstyle\sum_{i=1}^{L}\left[\tfrac{1+\sigma_i^z}{2}\Phi(m^z-2/L) + \tfrac{1-\sigma_i^z}{2}\Phi(m^z+2/L)\right] = E\Phi(m^z). \tag{3.104}$$

This means that the eigenvalue equation (3.101) for the ground state is reduced to another eigenvalue equation that has much smaller size $(L+1)$ than before (2^L):

$$\sum_{m' \in \mathcal{M}} H_{m,m'} \Phi(m') = \frac{E}{L} \Phi(m), \tag{3.105}$$

with $\mathcal{M} = \{-1, -1+2/L, \ldots, 1-2/L, 1\}$ and

$$H_{m,m'} = -(m)^p \delta_{m,m'} - \Gamma \left[\frac{1+m}{2} \delta_{m-2/L,m'} + \frac{1-m}{2} \delta_{m+2/L,m'} \right]. \tag{3.106}$$

We note that the matrix $H_{m,m'}$ is not symmetric even though the Hamiltonian $\hat{H}$ is Hermitian.

We denote the number of the state by $p(m)$:

$$p(m) \equiv \sum_{\vec{\sigma}} \delta_{m^z(\vec{\sigma}),m} = \frac{L!}{((1+m)L/2)!((1-m)L/2)!} \frac{1}{2^L}. \tag{3.107}$$

With this function, the expected value of a function $g(\hat{m}^z)$ in the ground state is calculated as

$$\frac{\langle \Phi | g(\hat{m}^z) | \Phi \rangle}{\langle \Phi | \Phi \rangle} = \sum_m g(m) p^\Gamma(m), \tag{3.108}$$

where we define a distribution function $p^\Gamma(m)$ as

$$p^\Gamma(m) = \frac{p(m)|\Phi(m)|^2}{\sum_m p(m)|\Phi(m)|^2}. \tag{3.109}$$

As the function $g(\hat{m}^z)$, we especially focus on the expected value and the variance of $\hat{m}^z$, which are denoted by $m(\Gamma)$ and $\sigma(\Gamma)$:

$$m(\Gamma) = \sum_m m p^\Gamma(m), \tag{3.110}$$

$$\sigma(\Gamma) = L \sum_m (m - m(\Gamma))^2 p^\Gamma(m). \tag{3.111}$$

It is well known that the ground state of quantum system is characterised by a variational principle:

$$E = \min_{|\Psi\rangle} \frac{\langle \Psi | \hat{H} | \Psi \rangle}{\langle \Psi | \Psi \rangle}, \tag{3.112}$$

where the optimal $|\Psi\rangle$ is reached at the ground state $|\Phi\rangle$. Then, from the same calculation for deriving (3.105), we rewrite (3.112) as

$$\frac{E}{L} = \min_{\tilde{\Phi}>0} \sum_{m} \tilde{p}(m) \sum_{m'} \tilde{\Phi}(m)^{-1} H_{m,m'} \tilde{\Phi}(m'), \tag{3.113}$$

where $\tilde{p}(m)$ is defined as

$$\tilde{p}(m) = \frac{p(m)\tilde{\Phi}(m)^2}{\sum_m p(m)\tilde{\Phi}(m)^2}. \tag{3.114}$$

We note that the optimal $\tilde{p}$ is equal to the ground state distribution function p^{Γ}.

Now, we can see the clear correspondence between this system and the one in the previous section. The eigenvalue equation (3.105) corresponds to (3.20). By using the eigenvector of (3.105), the distribution function (3.109) is defined, which structure corresponds to the distribution function in the biased ensemble (3.34). The transverse field corresponds to the biasing field s. The magnetisation and susceptibility directly corresponds to (3.38) and (3.39). Finally, the variational principle (3.113) corresponds to (2.29). The quantum phase transition takes place as the first order phase transition of $m(\Gamma)$ and $\sigma(\Gamma)$ as shown below. Here, we ask if we can apply to this system the same formulation as the one in the previous section.

Some Basics About the Quantum Phase Transition

Before going to the finite size scaling, here, we introduce some basics of the quantum phase transition in this model.

In the variational principle (3.113), we assume the large deviation principle of $\tilde{p}(m)$ for the ground state:

$$\tilde{p}(m) \sim \mathrm{e}^{-L\tilde{f}(m)} \tag{3.115}$$

with a large deviation function $\tilde{f}(m)$. Here, we recall that the number of the state $p(m)$ also satisfies the large deviation principle:

$$p(m) \sim \exp\left[-L\left(\frac{1+m}{2}\log(1+m) + \frac{1-m}{2}\log(1-m)\right)\right], \tag{3.116}$$

which indicates that $\tilde{\Phi}(m)$ also satisfies the large deviation scaling:

$$\tilde{\Phi} \sim \mathrm{e}^{-L\tilde{\phi}(m)/2} \tag{3.117}$$

with a large deviation function $\tilde{\phi}(m)$ defined as

$$\tilde{\phi}(m) = \tilde{f}(m) - \left[\frac{1+m}{2}\log(1+m) + \frac{1-m}{2}\log(1-m)\right]. \tag{3.118}$$

The saddle point equation for the sum of m in (3.113) is

$$\frac{\partial \tilde{f}(m)}{\partial m} = 0, \tag{3.119}$$

which is equivalent to

$$\frac{\partial \tilde{\phi}(m)}{\partial m} = \frac{1}{2} \log \frac{1+m}{1-m} \tag{3.120}$$

from (3.118). With these relations, by performing the saddle point approximation in the variational principle (3.113), we obtain a simple formula

$$\frac{E}{L} = \min_{m} \left[-m^p - \Gamma\sqrt{1-m^2} \right]. \tag{3.121}$$

This variational formula is a basics for this quantum phase transition and is well-known as shown in Ref. [11], for example. By solving (3.121), we obtain an equation for determining the expected value of the magnetization, which we denote by m^*:

$$m^*\Gamma = p\left(m^*\right)^{p-1}\sqrt{1-(m^*)^2}. \tag{3.122}$$

Since we are considering the system of $p \geq 3$, the system has the first order phase transition as follows. We denote the special value of Γ by $\Gamma_{\mathrm{c}}^{\infty}$, where the phase transition takes place. The phase transition is detected as the two solutions in the variational principle (3.121). Those two solutions correspond to paramagnetic solution and ferromagnetic solution. We thus denote them by $m_{\infty}^{\mathrm{pa}}(=0)$ and m_{∞}^{fe}, respectively. In order to determine $\Gamma_{\mathrm{c}}^{\infty}$ and m_{∞}^{fe}, we use a condition of first order phase transition:

$$\left[-m^p - \Gamma_{\mathrm{c}}^{\infty}\sqrt{1-m^2} \right]\Bigg|_{m_{\infty}^{\mathrm{pa}}=0} = \left[-m^p - \Gamma_{\mathrm{c}}^{\infty}\sqrt{1-m^2} \right]\Bigg|_{m=m_{\infty}^{\mathrm{fe}}} \tag{3.123}$$

and the relation of m^* and Γ as shown in (3.122).

We numerically solve the variational principle (3.121) for $p = 3$, and plot the obtained E/L and m in Fig. 3.13. On the same figure, we also plot the numerical examples of E/L, $m(\Gamma)$, and $\sigma(\Gamma)$ obtained from diagonalising the matrix (3.106) for finite-size systems. The first order phase transition is observed around $\Gamma_{\mathrm{c}}^{\infty} \simeq 1.3$ in the figure. We also observe the finite size correction of m and σ in it. In the following section, we approach to this finite size structure, where the problem doesn't allow us to utilise the naive perturbation approach from the formulation explained here. In order to obtain the finite size correction, we need to use the same formulation developed in the previous section for the KCM.

3.3.2 Finite-Size Scaling

Definition of Scaling Function

First, we define the transition point for finite-size system L, Γ_{c}^{L}, as

$$\Gamma_{\mathrm{c}}^{L} = \mathrm{Argmax}_{\Gamma}\sigma(\Gamma). \tag{3.124}$$

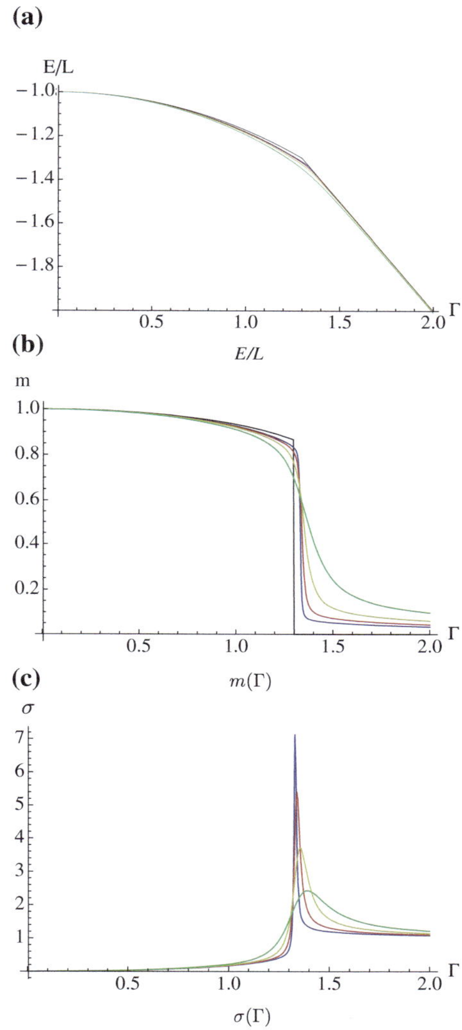

Fig. 3.13 Numerical examples of the ground state energy E/L (**a**), the magnetisation $m(\Gamma)$ (**b**), and the susceptibility $\sigma(\Gamma)$ (**c**). *Blue*, *red*, *yellow*, and *green lines* correspond to the system size $L = 50$, $L = 40$, $L = 30$, and $L = 20$, respectively. The black lines are obtained from the variational principle (3.121), which represents the result in $L \to \infty$ limit. © IOP Publishing. Reproduced with permission. All rights reserved

Then, we quantify the width of the first order phase transition point by using the derivative of $m(\Gamma_c^L)$:

$$\kappa \equiv -\partial m(\Gamma_c^L)/\partial\Gamma, \tag{3.125}$$

which is inversely proportional to the width. With this κ, we introduce scaling functions by

$$\tilde{m}(x) = m(\Gamma_c^L + x\kappa^{-1}), \tag{3.126}$$

and

$$\tilde{\sigma}(x) = \frac{\sigma(\Gamma_c^L + x\kappa^{-1})}{\sigma(\Gamma_c^L)}. \tag{3.127}$$

Application of Our Method to Quantum Phase Transition—Scaling Function

Here, to quantum phase transition, we apply our method developed in the previous section (Sect. 3.2.3). Before applying our method, we divide the distribution function at the coexistence point $p^{\Gamma_c^L}(m)$ into two regions:

$$p^{\Gamma_c^L}(m) = \delta_{m\le m_c} P_p(m) + \delta_{m>m_c} P_f(m), \tag{3.128}$$

where m_c is the boundary of these two regions, which is defined as

$$\sum_{m\le m_c} P_p(m) = \sum_{m>m_c} P_f(m) = 1/2. \tag{3.129}$$

The region $m \le m_c$ corresponds to paramagnet region, whereas, the region $m > m_c$ corresponds to ferromagnet region.

Now, around this coexistence point, we assume that the distribution function is written by using a mixing function $a^*(\Gamma)$:

$$p^{\Gamma}(m) = (1 + a^*(\Gamma))\delta_{m\le m_c} P_p(m) + (1 - a^*(\Gamma))\delta_{m>m_c} P_f(m). \tag{3.130}$$

We note that the normalization condition is satisfied thanks to (3.129). The mixing function $a^*(\Gamma)$ is determined by the variational principle (3.113) as follows. From (3.109) and (3.130), we have $\Phi(m)$ in terms of the mixing function $a^*(\Gamma)$:

$$\Phi(m) \propto \sqrt{(1 + a^*(\Gamma))\frac{P_p(m)}{p(m)}}\delta_{m\le m_c} + \sqrt{(1 - a^*(\Gamma))\frac{P_f(m)}{p(m)}}\delta_{m>m_c}. \tag{3.131}$$

We substite (3.130) and (3.131) into the variational principle (3.113), and maximise it with respect to $a^*(\Gamma)$. The obtained maximiser is the desired mixing function. Since the calculation is almost the same as the one in the Sect. 3.2.3, we don't repeat it. We only write down the result here.

$$a^*(\Gamma) = \frac{x\left[\langle m\rangle_p - \langle m\rangle_f\right]^{-1}}{\sqrt{1 + 4x^2\left[\langle m\rangle_p - \langle m\rangle_f\right]^{-2}}} + O(\kappa^{-1}). \tag{3.132}$$

with $x = \kappa(\Gamma - \Gamma_c)$.

Once we obtain the distribution function around the coexistence region, the scaling function of $m(\Gamma)$, $\sigma(\Gamma)$ directly follows. Indeed, $m(\Gamma)$ and $\sigma(\Gamma)$ are written in terms of the distribution function as

$$m(\Gamma) = \frac{\langle m\rangle_p}{2}\left[1 + a^*(\Gamma)\right] + \frac{\langle m\rangle_f}{2}\left[1 - a^*(\Gamma)\right], \tag{3.133}$$

$$\sigma(\Gamma) = L\left\{\frac{\langle m^2\rangle_{\mathrm{p}}}{2}\left[1 + a^*(\Gamma)\right] + \frac{\langle m^2\rangle_{\mathrm{f}}}{2}\left[1 - a^*(\Gamma)\right] - m(\Gamma)^2\right\}, \tag{3.134}$$

where $\langle\ \rangle_{\mathrm{p}}$ and $\langle\ \rangle_{\mathrm{f}}$ are defined as $\langle g\rangle_{\mathrm{p}} = 2\sum_{m\le m_{\mathrm{c}}} P_{\mathrm{p}}(m)g(m)$ and $\langle g\rangle_{\mathrm{f}} = 2\sum_{m>m_{\mathrm{c}}} P_{\mathrm{f}}(m)g(m)$, which corresponds to the expected values in the paramagnetic phase and the ferromagnetic phase, respectively. Then, from the definition of the scaling functions (3.126) and (3.127), we obtain

$$\tilde{m}(x) = \frac{1}{2}\left[\langle m\rangle_{\mathrm{p}} + \langle m\rangle_{\mathrm{f}} - \frac{2x}{\sqrt{1 + 4x^2\left[\langle m\rangle_{\mathrm{p}} - \langle m\rangle_{\mathrm{f}}\right]^{-2}}}\right], \tag{3.135}$$

and

$$\tilde{\sigma}(x) = \frac{1}{C}\left[\langle m^2\rangle_{\mathrm{p}} + \langle m^2\rangle_{\mathrm{f}} - \frac{2x\left[\langle m^2\rangle_{\mathrm{p}} - \langle m^2\rangle_{\mathrm{f}}\right]\left[\langle m\rangle_{\mathrm{p}} - \langle m\rangle_{\mathrm{f}}\right]^{-1}}{\sqrt{1+4x^2\left[\langle m\rangle_{\mathrm{p}} - \langle m\rangle_{\mathrm{f}}\right]^{-2}}} - 2\tilde{m}(x)^2\right], \tag{3.136}$$

where C is

$$C = \langle m^2\rangle_{\mathrm{p}} + \langle m^2\rangle_{\mathrm{f}} - 2\tilde{m}(0)^2. \tag{3.137}$$

Finally, by taking $L \to \infty$ limit and noticing

$$\lim_{L\to\infty}\langle m\rangle_{\mathrm{p}} = \lim_{L\to\infty}\langle m^2\rangle_{\mathrm{p}} = 0, \tag{3.138}$$

$$\lim_{L\to\infty}\langle m\rangle_{\mathrm{f}} = m_{\infty}^{\mathrm{fe}}, \tag{3.139}$$

and

$$\lim_{L\to\infty}\langle m^2\rangle_{\mathrm{f}} = (m_{\infty}^{\mathrm{fe}})^2, \tag{3.140}$$

we arrive at

$$\tilde{m}_{\infty}(x) = \lim_{L\to\infty}\tilde{m}(x) = \frac{1}{2}\left[m_{\infty}^{\mathrm{fe}} - \frac{2x}{\sqrt{1 + 4x^2(m_{\infty}^{\mathrm{fe}})^{-2}}}\right], \tag{3.141}$$

$$\tilde{\sigma}_{\infty}(x) = \lim_{L\to\infty}\tilde{\sigma}(x) = \frac{(m_{\infty}^{\mathrm{fe}})^2}{(m_{\infty}^{\mathrm{fe}})^2 + 4x^2}. \tag{3.142}$$

(3.135), (3.136) have the same form as (3.87), (3.88) , and also (3.141), (3.142) have the same as (3.89), (3.90). We numerically check (3.135) and (3.136). For fixed finite L, we numerically evaluate $\langle m\rangle_{\mathrm{p}}$, $\langle m^2\rangle_{\mathrm{p}}$, $\langle m\rangle_{\mathrm{f}}$, and $\langle m^2\rangle_{\mathrm{f}}$. Then, we plot (3.135) and

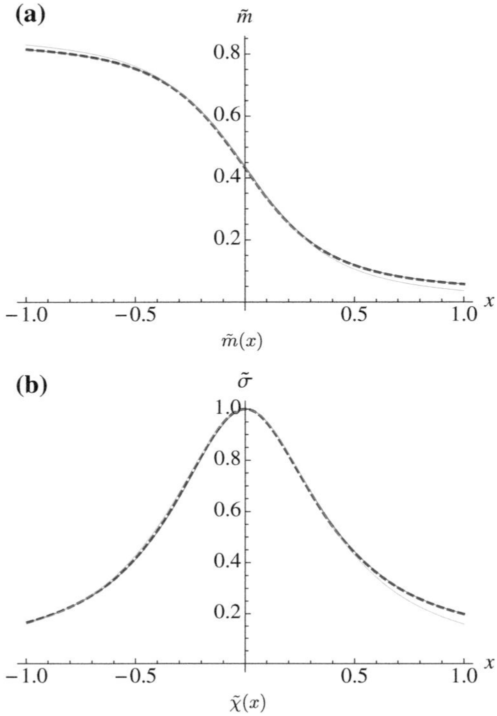

Fig. 3.14 Numerical check of the analytical expressions of the rescaled magnetisation $\tilde{m}(x)$ (**a**) and the susceptibility $\tilde{\chi}(x)$ (**b**). We set $p = 3$ and $L = 100$. The *red solid lines* are the analytical results obtained from (3.135) and (3.136). In order to use the formulas, we need the expected values of magnetisation only at $x = 0$ ($\Gamma = \Gamma_c^L$). For this, we diagonalise the corresponding eigenvalue problem at only that point. The *blue dashed lines* are the numerical results obtained from diagonalising the eigenvalue problem for every x (or Γ). On the same figure, for a comparison, we also plot the infinite-size scaling functions (3.141) and (3.142) with *yellow solid lines*.

(3.136) in Fig. 3.14. On the same figure, we also plot the scaling function obtained from solving the largest eigenvalue problem (3.105) for each s with the same system size L. We can see the good agreement of them. Finally, for seeing the speed of convergence to the infinite-size scaling functions (3.141) and (3.142), we plot them in Fig. 3.14 with yellow lines. We understand that larger system sizes are required for observing the convergences to them.

Application of Our Method to Quantum Phase Transition—Scaling Factor and Exponentially Small Gap

Here, to quantum phase transition, we apply our formula for obtaining the width of the coexistence region. Then, we find that the obtained expression is exactly the same as the one derived by Bapst and Semerjian for giving the exponentially small energy gap between the ground state and the first excited state when the quantum phase transition takes place [11].

We first denote the free energy density for the ground state at the transition point Γ_c^∞ by $f_{\Gamma_c^\infty}(m)$:

$$f_{\Gamma_c^\infty}(m) = -\lim_{L\to\infty} \frac{1}{L} \log p^{\Gamma_c^\infty}(m). \tag{3.143}$$

Then, by applying our formulation developed in Sect. 3.2.3, we obtain the same formula as (3.85). That is,

$$\lim_{L\to\infty}\frac{1}{L}\log\kappa = f_{\Gamma_c^\infty}(m_c^\infty), \tag{3.144}$$

where we denote correcting point between two phases in infinite size limit m_c^∞ by

$$m_c^\infty \equiv \lim_{L\to\infty} m_c. \tag{3.145}$$

Next, we determine the free energy. From (3.105), the eigenvalue equation of ground state at $\Gamma = \Gamma_c^\infty$ is

$$-m^p - \Gamma_c^\infty\left[\frac{1+m}{2}\frac{\Phi(m-2/L)}{\Phi(m)} + \frac{1-m}{2}\frac{\Phi(m+2/L)}{\Phi(m)}\right] = \frac{E}{L}. \tag{3.146}$$

In this equation, we assume a large deviation principle in the function $\Phi(m)$:

$$\Phi(m) = e^{-(L/2)g(m)}. \tag{3.147}$$

We evaluate the leading term of (3.146), which becomes

$$-m^p - \Gamma_c^\infty\left[\frac{1+m}{2}e^{\partial g/\partial m} + \frac{1-m}{2}e^{-\partial g/\partial m}\right] = e_c^\infty, \tag{3.148}$$

with the definition of e_c^∞ as $\lim_{L\to\infty} E/L|_{\Gamma=\Gamma_c^\infty} \equiv e_c^\infty$. There are two solutions of this differential equation. We denote them by $g_\pm(m)$. Those are given as

$$g_\pm(m) = \int_0^m d\tilde{m}\log\left[-\frac{\tilde{m}^p + e_c^\infty}{(1+\tilde{m})\Gamma_c^\infty} \pm \sqrt{\left(\frac{\tilde{m}^p + e_c^\infty}{(1+\tilde{m})\Gamma_c^\infty}\right)^2 - \frac{1-\tilde{m}}{1+\tilde{m}}}\right]. \tag{3.149}$$

We also denote the corresponding free energies to $g_\pm(m)$ by $f_\pm(m)$, which are given as

$$\begin{aligned} f_\pm(m) &\equiv g_\pm(m) + \frac{1+m}{2}\log(1+m) + \frac{1-m}{2}\log(1-m) \\ &= \int_0^m d\tilde{m}\log\left[-\frac{\tilde{m}^p + e_c^\infty}{\sqrt{1-\tilde{m}^2}\Gamma_c^\infty} \pm \sqrt{\left(\frac{\tilde{m}^p + e_c^\infty}{\sqrt{1-\tilde{m}^2}\Gamma_c^\infty}\right)^2 - 1}\right] \end{aligned} \tag{3.150}$$

The true free energy $f_{\Gamma_c^\infty}(m)$ is the combination of $f_+(m)$ and $f_-(m)$. For constructing it, we recall

$$f_{\Gamma_c^\infty}(0) = f_{\Gamma_c^\infty}(m_{\text{fe}}^\infty) = 0, \tag{3.151}$$

$$\left.\frac{\partial f_+(m)}{\partial m}\right|_{m=0} > 0, \tag{3.152}$$

and

$$\left.\frac{\partial f_-(m)}{\partial m}\right|_{m=0} < 0. \tag{3.153}$$

From these relations and an assumption that there is only one connecting point, the free energy $f_{\Gamma_c^\infty}(m)$ is uniquely determined as

$$f_{\Gamma_c^\infty}(m) = f_+(m) \tag{3.154}$$

for $m \le m_c^\infty$, and

$$f_{\Gamma_c^\infty}(m) = f_-(m) + \text{const.} \tag{3.155}$$

for $m > m_c^\infty$, where the constant and m_c^∞ is determined from (3.151) and the continuity condition

$$\lim_{m \to m_c+0} f_{\Gamma_c^\infty}(m) = \lim_{m \to m_c-0} f_{\Gamma_c^\infty}(m). \tag{3.156}$$

From the obtained free energy with the values of the parameters Γ_c^∞, e_c^∞, and m_{fe}^∞ determined from (3.121)–(3.123), one can calculate the gap given by (3.144) in principle. Here, however, we show that the gap can be expressed as a much simpler form with the aide of a relation

$$f_+(m) = -f_-(m) + \text{const.}, \tag{3.157}$$

which is confirmed by the direct substitution. Indeed, from this, we obtain

$$f_+(m_c^\infty) = \frac{1}{2} f_+(m_{fe}^\infty), \tag{3.158}$$

which leads to

$$\begin{aligned}
\lim_{L\to\infty} \frac{1}{L} \log \kappa &= \frac{1}{2} f_+(m_{fe}^\infty) \\
&= \int_0^{m_{fe}^\infty} dm \log\left[-\frac{m^p + e_c^\infty}{\sqrt{1 - m^2 \Gamma_c^\infty}} + \sqrt{\left(\frac{m^p + e_c^\infty}{\sqrt{1 - m^2\Gamma_c^\infty}}\right)^2 - 1}\right] \\
&= \int_0^{m_{fe}^\infty} dm \cosh^{-1}\left(\frac{m^p + e_c^\infty}{\sqrt{1 - m^2\Gamma_c^\infty}}\right),
\end{aligned} \tag{3.159}$$

where we used a basic mathematical fact that the following equations are equivalent: $\cosh x = A$ and $e^x = A \pm \sqrt{A^2 - 1}$. The expression of this formula is equivalent to the one for determining the exponentially small gap between the ground state energy and the first excited energy at the coexistence region derived by Bapst and Semerjian in Ref. [11]. This indicates that the width of the coexistence region and the energy gap are quantitatively equivalent.

3.4 Conclusions

In this chapter, we analysed a mean-field FA model. Especially, we focused on the stationary state corresponding to the biased ensemble (Sect. 2.2.5). Then, we derived an analytical expression of the static free energy at the transition point. The free energy itself had a singularity in the infinite system size limit, from which the domain of the variable of the system was divided into two regions naturally. Furthermore, we proposed a method to obtain an analytical expression of scaling functions around the phase transition point, where the method was an extension of the one by Borgs and Kotecký for equilibrium ferromagnets [8, 9]. For the extension, we combined the variational formula (2.29) with their methods to fill the gap between equilibrium phase transitions and dynamical phase transitions. Finally, by utilising the mathematical fact that dynamical phase transitions and quantum phase transitions are mathematically equivalent, we also derived the same scaling function in the mean-field p-spin model.

As a future problem, we here mention the application of our method to finite dimensional systems. We expect that our basic idea to derive scaling functions should be applied in any dimension. This basic idea is as follows: For a given distribution function at the coexistence region, we define the distribution function *around* the coexistence region as the sum of each phase (obtained from the distribution function at each coexistence phase) with multiplying each of them by a special coefficient, which is an unknown function of s. Then, these unknown coefficients are determined by the variational principle (2.29). Since the method by Borgs and Kotecký [8, 9] is generally true even for finite dimensional systems, this expectation is quite legitimate. Indeed, soon after the submission of our work, Campostrini and collaborators presented some results independently about scaling functions of quantum first order phase transition for a finite dimensional case [15]. Interestingly, the scaling functions take the form similar to ours. They derived these results based on an approach with a two-level effective model. Yet, the connection between ours and theirs is not clarified. Further studies are needed.

Apart from the scaling functions, it is also interesting to see the modified free energies at the coexistence region in finite dimensional cases. In the case of mean-field model, the modified free energy itself has a singularity, from which the domain is divided into two regions. The next question may be whether there is a similar structure in finite dimensional cases. Since it is difficult to obtain analytical results in finite dimensional cases in general, we will need to rely on some numerical simulations. Furthermore, it will not be a straightforward problem to know which type of free energy (large deviation function) shows interesting behaviour. However, we believe that it is worth to challenge this research, since, if we success in this research, it may be possible to connect quantitatively the dynamical phase transitions with the origin of glassy features such as dynamical heterogeneities.

References

1. J.P. Garrahan, R.L. Jack, V. Lecomte, E. Pitard, K. van Duijvendijk, F. van Wijland, Phys. Rev. Lett. **98**, 195702 (2007)
2. J.P. Garrahan, R.L. Jack, V. Lecomte, E. Pitard, K. van Duijvendijk, F. van Wijland, J. Phys. A **42**, 075007 (2009)
3. J.P. Garrahan, P. Sollich, C. Toninelli, (2010) arXiv:1009.6113, which is a chapter in Ref. (L. Berthier, G. Biroli, J.-P. Bouchaud, L. Cipelletti, W. van Saarloos, (eds.), *Dynamical Heterogeneities in Glasses, Colloids, and Granular Media* (Oxford University Press, Oxford, 2011))
4. T. Bodineau, C. Toninelli, Commun. Math. Phys. **311**, 357 (2012)
5. T. Bodineau, V. Lecomte, C. Toninelli, J. Stat. Phys. **147**, 1 (2012)
6. R.L. Jack, P. Sollich, J. Phys. A **47**, 015003 (2014)
7. T. Nemoto, S. Sasa, Phys. Rev. Lett. **112**, 090602 (2014)
8. C. Borgs, R. Kotecký, J. Stat. Phys. **61**, 79 (1990)
9. C. Borgs, R. Kotecký, Phys. Rev. Lett. **68**, 1734 (1992)
10. T. Jörg, F. Krzakala, J. Kurchan, A.C. Maggs, J. Pujos, Europhys. Lett. **89**, 40004 (2010)
11. V. Bapst, G. Semerjian, J. Stat. Mech. (2012) P06007
12. T. Nemoto, V. Lecomte, S. Sasa, F. van Wijland, J. Stat. Mech. (2014) P10001
13. C. Giardina, J. Kurchan, L. Peliti, Phys. Rev. Lett. **96**, 120603 (2006)
14. C. Giardina, J. Kurchan, V. Lecomte, J. Tailleur, J. Stat. Phys. **145**, 787 (2011)
15. M. Campostrini, J. Nespolo, A. Pelissetto, E. Vicari, Phys. Rev. Lett. **113**, 070402 (2014)

Chapter 4
van Zon–Cohen Singularity and a Negative Inverse Temperature

4.1 Introduction

A symmetry property of the fluctuation of entropy production was found in 1993. This property was called the fluctuation theorem [1], which took the form of the extension of the linear response theory, and also the well-known second law of thermodynamics [2–6]. The property included the higher order fluctuation, so that the experimental verification of the theorem required small size systems described by Stochastic thermodynamics [7]. Wang et al. were the first group that verified the fluctuation theorem in a real experiment in 2001 [8]. For this experiment, a Brownian particle dragged by an optical tweezer was studied, and the fluctuation theorem for the work done by the tweezer was confirmed. See also Ref. [9] for the detailed analysis of the system that they studied.

When we consider the higher order fluctuation of heat or work, the relationship between these two fundamental quantities are not trivial. For the expectation value of these quantities are equivalent (in reverse sign). This is the first law of thermodynamics. However, we can no longer rely on this law for higher order moments. In the case of the Brownian particle dragged by an optical tweezer, this relation was explicitly studied by van Zon and Cohen [10, 11], where the heat was defined as the energy dissipated by the particle. In one hand, they derived the fluctuation theorem for the work, but on the other hand, they derived another fluctuation theorem for the heat. This new fluctuation theorem has a different form from the one that had been known, and they called this property, *an extended fluctuation theorem*. After this discovery, several studies to clarify this new symmetry have been done. See Refs. [12–19], for example.

For explaining the extended fluctuation theorem, we here consider a cumulant generating function,

T. Nemoto, *Phenomenological Structure for the Large Deviation Principle in Time-Series Statistics*, Springer Theses,
DOI 10.1007/978-981-287-811-3_4

$$G(h) = \frac{1}{t} \log \left\langle e^{hQ(t)} \right\rangle, \tag{4.1}$$

and a biased distribution function,

$$P_h(x, t) = e^{-tG(h)} \left\langle \delta(x(t) - x) e^{hQ(t)} \right\rangle, \tag{4.2}$$

where $x(t)$ is the position of the particle at time t, $Q(t)$ is the accumulated heat from time $t = 0$ to $t = t$, the parameter h is a biasing field, and $P_h(x, t)$ is the expectation value of $\delta(x(t) - x)$ with respect to the path probability density biased by an exponential factor $e^{hQ(t) - tG(h)}$. It means that $P_h(x, t)$ is the distribution function of $x(t)$ in the exponentially biased ensemble introduced as (2.11). The rare trajectories characterised by the large deviation principle of $Q(t)$ has larger weight in this biased ensemble for $h > 0$, and smaller weight for $h < 0$. As reported in the original papers [10, 11], the extended fluctuation theorem appears with a singularity of the cumulant generating function $G(h)$. More precisely, $G(h)$ becomes singular, when we set $|h|$ to be a value larger than a special value h_c. Here in this chapter, we call this singularity a van Zon–Cohen singularity.

van Zon–Cohen singularity is related to rare trajectories, since this singularity emerges when the biasing field is larger than h_c. However, the relationship between this singularity and the behaviour of the particle in rare trajectories is not clarified yet. Recently, the significance of fluctuation in non equilibrium physics has been recognized, as explained in the introduction of this thesis. Thus, it is important to study this singularity with a systematic method and clarify the relationship between them, and furthermore, to seek for the same kind of singularity for the other quantities than heat. In this chapter, we tackle this problem.

We consider a Brownian particle on a moving periodic potential. The model is the over-damped version of the one studied by Lebowitz and Spohn [6]. We systematically analyse this system by using a boundary layer analysis, and then we calculate the cumulant generating function and the corresponding biased ensemble of heat and work. We find that the biased distribution function becomes a canonical distribution function in which the inverse temperature is modified by h, when the period and the depth of the potential are both large. Then, since the inverse temperature is modified by the parameter, when $|h| > h_c$, the inverse temperature becomes negative and the two limiting operations, which are the trapping particle limit and the large observation time limit, become non-interchangeable. This non-interchangeability corresponds to the van Zon–Cohen singularity. Furthermore, by checking the conditional distribution function given $Q(t)/t = q$, we see that how hard to observe the trajectories causing the singularity is. These discussions indicate that the similar kind of singularities may exist in the other quantities.

The organization of this chapter is the following. In Sect. 4.2, we define the model we study and have a brief introduction of the biased process. In Sects. 4.3 and 4.4, we explain our main results and then, we derive them. Finally, we make concluding remarks in Sect. 4.5. The Boltzmann constant k_B is set to be 1 throughout this chapter. The discussion in this chapter is based on our published paper [20].

4.2 Set up

4.2.1 Model

A one-dimensional Brownian particle is considered. We denote the temperature of the solvent by T and the position of the particle by $x(t) \in \mathbb{R}$. We exert a force $-\partial U(x)/\partial x$ on the particle, where $U(x)$ is a periodic potential. The period of the potential is denoted by $2L$, which means that $U(x)$ satisfies

$$U(x) = U(x + 2nL) \tag{4.3}$$

for $n = \pm 1, \pm 2, \ldots$. Then, we move the potential toward the negative direction of x with a constant velocity v. We mention that adding force in this manner in real experiments is possible, using an optical tweezer [21]. In this setting, the motion of the particle is described by the Langevin equation

$$\dot{x}(t) = -\frac{1}{\gamma}\frac{\partial}{\partial y}U(y)\bigg|_{y=x(t)+vt} + \sqrt{\frac{2T}{\gamma}}\xi(t), \tag{4.4}$$

where γ is a friction constant, and $\xi(t)$ is the Gaussian white noise satisfying $\langle\xi(t)\rangle = 0$ and $\langle\xi(t)\xi(s)\rangle = \delta(t - s)$.

For making the analysis easier, we define a new variable $y(t)$ as the position of the particle measured by the reference frame, which is moving with the periodic potential. In other words, $y(t)$ is defined as

$$y(t) \equiv x(t) + vt - 2nL, \tag{4.5}$$

where n is an integer determined by $-L \leq x(t) + vt - 2nL < L$. Here, we note that $y(t)$ is confined to $[-L, L)$. We also have the Langevin equation for $y(t)$, which is obtained from (4.4). That is written as

$$\dot{y}(t) = -\frac{1}{\gamma}\frac{\partial}{\partial y}U\left(y(t)\right) + v + \sqrt{\frac{2T}{\gamma}}\xi(t). \tag{4.6}$$

Recently, this type of systems has been used for experimental tests of several non-equilibrium relations [22–24].

Then, we specify the condition of the potential that we study. Since we consider the trapped particle limit, we consider $U(x)$ satisfying the condition

$$\lim_{L\to\infty} \frac{\partial U(y)}{\partial y}\bigg|_{y=YL} = \infty \tag{4.7}$$

for $0 < |Y| \leq 1$. For example, the harmonic potential

$$U_{\text{harmo}}(x) = \frac{1}{2}k(x - 2nL)^2 \tag{4.8}$$

and the quartic potential

$$U_{\text{quart}}(x) = \frac{1}{4}k_4(x - 2nL)^4, \tag{4.9}$$

with an integer n determined by

$$-L \le x - 2nL < L, \tag{4.10}$$

satisfy this condition (4.7). However, we mention that the linear potential

$$U_{\text{linear}}(x) = k_1|x - 2nL| \tag{4.11}$$

does not satisfy the condition (4.7). We show the numerical example of the harmonic potential (4.8) in Fig. 4.1.

Next, we specify the quantity we consider the work $W(t)$ done by the periodic potential and the heat $Q(t)$ dissipated by the particle are considered. They are calculated from a particle trajectory as follows: Since the periodic potential exerts the force $-\partial U(y)/\partial y|_{y=x(t)+vt}$ on the particle and it moves with the constant velocity $-v$, $\dot{W}(t)$ is calculated as

$$\dot{W}(t) = (-v)\left[-\frac{\partial}{\partial y}U(y)\bigg|_{y=x(t)+vt}\right]. \tag{4.12}$$

Also, according to Sekimoto's argument [7], the rate of the heat dissipation $\dot{Q}(t)$ is evaluated as

$$\dot{Q}(t) = \dot{x} \circ \left[\gamma\dot{x} - \sqrt{2\gamma T}\xi(t)\right], \tag{4.13}$$

where the multiplication $\circ$ represents the Stratonovich interpretation [25]. We note that the first law of thermodynamics is satisfied:

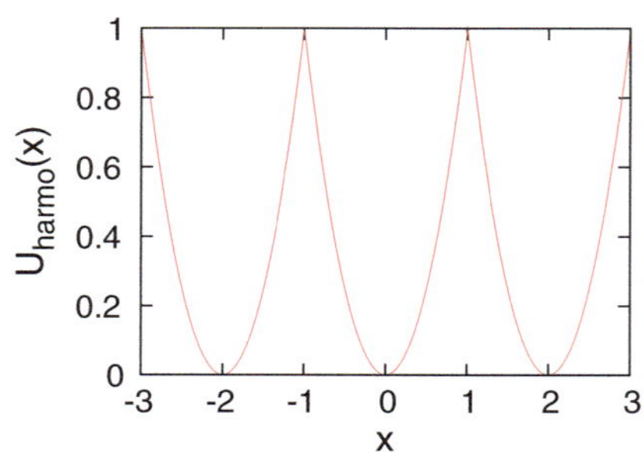

Fig. 4.1 The harmonic potential (4.8) with $k = 2$ and $L = 1$. This is an example of the potential that satisfies the condition (4.7). Reprinted with permission from Ref. [20]. Copyright 2012 by American Physical Society

$$\int_{t_1}^{t_2} dt \left(\dot{W}(t) - \dot{Q}(t)\right) = U(x(t_2) + vt_2) - U(x(t_1) + vt_1), \tag{4.14}$$

and that these $\dot{Q}(t)$ and $\dot{W}(t)$ can be expressed as the function of $y(t)$,

$$\dot{W}(t) = v \frac{\partial}{\partial y} U\,(y(t))\,, \tag{4.15}$$

$$\dot{Q}(t) = \frac{1}{\gamma} \left(\frac{\partial U(y)}{\partial y} \right)^2 - \sqrt{\frac{2T}{\gamma}} \frac{\partial U(y)}{\partial y} \circ \xi(t). \tag{4.16}$$

The expectation value with respect to the realisation of the noise $(\xi(s))_{s=0}^{\infty}$ with an initial distribution function $p(y_0) = \langle \delta(y(0) - y_0) \rangle$ is denoted by $\langle\,\rangle_p$. With this notation, we define a joint distribution function $P(y_0, y, t|p)$ as

$$P(y_0, y, t|p) = \langle \delta(y(0) - y_0) \delta(y(t) - y) \rangle_p\,. \tag{4.17}$$

4.2.2 Biased Process and Cumulant Generating Functions

Here, we briefly introduce a biased process. For the detailed introduction, see the Sect. 2.2 in this thesis. We denote by $f(t)$ a function of time, which depends on the trajectory of the particle $(y(s))_{s=0}^{t}$. We define the expectation value of $f(t)$ with respect to a biased process as

$$\langle f(t) \rangle_p^{h_w, h_q} \equiv e^{-tG(h_w, h_q, t|p)} \left\langle f(t) e^{W(t)h_w + Q(t)h_q} \right\rangle_p, \tag{4.18}$$

where $G(h_w, h_q, t|p)$ is a cumulant generating function

$$G(h_w, h_q, t|p) = \frac{1}{t} \log \left\langle e^{W(t)h_w + Q(t)h_q} \right\rangle_p. \tag{4.19}$$

The cumulant generating function corresponds to a thermodynamic free energy in terms of thermodynamic formalism [26]. The two parameters h_w and h_q are called biasing fields. The biased expectation value $\langle f(t) \rangle_p^{h_w, h_q}$ becomes back to the original expectation value $\langle f(t) \rangle_p$ when we set $h_w = h_q = 0$, and also the cumulant generating function $G(h_w, h_q, t|p)$ becomes 0 in the same setting $h_w = h_q = 0$. Finally with (4.18), we define a biased joint distribution function as

$$P_{h_w, h_q}(y_0, y, t|p) = \langle \delta(y(t) - y) \delta(y(0) - y_0) \rangle_p^{h_w, h_q}\,. \tag{4.20}$$

Next, we define two functions that help us to analyse the asymptotic behaviour of $G(h_w, h_q, t|p)$ in $t \to \infty$. The first one is the limiting value of $G(h_w, h_q, t|p)$ in $t \to \infty$,

$$G_{\text{scaled}}(h_w, h_q) = \lim_{t\to\infty} G(h_w, h_q, t|p), \tag{4.21}$$

which is called a scaled cumulant generating function [27]. The second function is the one representing the deviation of $G(h_w, h_q, t|p)$ from that limiting value,

$$H_{\text{ex}}(h_w, h_q|p) = \lim_{t\to\infty} t\left[G(h_w, h_q, t|p) - G_{\text{scaled}}(h_w, h_q)\right], \tag{4.22}$$

which is called an excess quantity of the cumulant generating function. We note that this excess cumulant generating function was used for the calculation of an excess heat in Ref. [28]. We also note that we explicitly indicated the initial distribution function dependence in (4.22) as p, because the excess cumulant generating function depends on it in general. By using these two functions, we can express $G(h_w, h_q, t|p)$ as

$$G(h_w, h_q, t|p) \simeq G_{\text{scaled}}(h_w, h_q) + \frac{1}{t} H_{\text{ex}}(h_w, h_q|p). \tag{4.23}$$

We mention that the difference of the left-hand side and the right-hand side of (4.23) is $O(e^{-at})$, where a is a positive constant, as shown (4.67) below.

4.2.3 Biased Distribution Function and Conditional Distribution Function

Here, let us show a useful relationship between the biased distribution function and a conditional distribution function for a fixed heat.

A joint distribution function of $y(0)$, $y(t)$, and $Q(t)/t$, which is defined by

$$P(y_0, y, q, t|p) \equiv \langle \delta(y(0) - y_0)\delta(y(t) - y)\delta(Q(t)/t - q)\rangle_p, \tag{4.24}$$

is considered. We use this definition in the right-hand side of (4.20). Then, we obtain

$$\log P_{0,h_q}(y_0, y, t|p) + tG(0, h_q, t|p) = \log \int dq P(y_0, y, q, t|p) e^{tqh_q}. \tag{4.25}$$

Next, we introduce a function $I(y_0, y, q, t|p)$, which corresponds to a large deviation function in $t \to \infty$ limit, as

$$I(y_0, y, q, t|p) \equiv -\frac{1}{t} \log P(y_0, y, q, t|p). \tag{4.26}$$

Since the large deviation principle is satisfied, we assume the following asymptotic form for $I(y_0, y, q, t|p)$,

$$I(y_0, y, q, t|p) = I_0(q) + \frac{1}{t} I_1(y_0, y, q|p) + o(1/t) \tag{4.27}$$

for large t. Here, $I_0(q)$ is the large deviation function of $Q(t)/t$. We evaluate the right-hand side of (4.25) by using this asymptotic form, with a saddle point method. We then obtain

$$t \max_q \left[h_q q - I_0(q)\right] - I_1(y_0, y, q^*|p) + \frac{1}{1/t} o(1/t), \tag{4.28}$$

where q^* is defined as

$$q^* \equiv \underset{q}{argmax} \left[h_q q - I_0(q)\right]. \tag{4.29}$$

From (4.23), (4.25) and (4.28), the scaled cumulant generating function $G_{\text{scaled}}(0, h_q)$ and the excess cumulant generating function $H_{\text{ex}}(0, h_q|p)$ are connected with these quantities,

$$G_{\text{scaled}}(0, h_q) = \max_q \left[h_q q - I_0(q)\right], \tag{4.30}$$

and

$$\log P_{0,h_q}(y_0, y, t|p) + H_{\text{ex}}(0, h_q|p) = -I_1(y_0, y, q^*|p) + \frac{1}{1/t} o(1/t). \tag{4.31}$$

Here, we note that (4.30) is a well-known relation between a large deviation function and a scaled cumulant generating function [27, 29]. Finally, by combining (4.31) with (4.27) and using the normalization condition for $P_{0,h_q}(y_0, y, t|p)$, we arrive at

$$P_{0,h_q}(y_0, y, t|p) = \frac{P(y_0, y, q^*, t|p)}{P(q^*, t|p)} + \frac{1}{1/t} o(1/t), \tag{4.32}$$

where $P(q^*, t|p)$ is a normalization constant defined by

$$P(q^*, t|p) \equiv \int dy_0 dy P(y_0, y, q^*, t|p). \tag{4.33}$$

This equation means that the biased distribution function is equivalent to the conditional distribution function of $y(0)$ and $y(t)$ for given $Q(t)/t = q^*$.

4.3 Results

In this section, we explain our results. The derivation of them will be developed in the following section.

The stationary distribution function of y is denoted by $p_{\text{st}}^{U,v,\beta}(y)$, where the superscripts U, v and β indicate the shape of the periodic potential, the velocity of the moving potential, and the inverse temperature of the solvent, respectively. We then define a canonical distribution function $p_{\text{can}}^{U,v,\beta}(y)$ as

$$p_{\text{can}}^{U,v,\beta}(y) \equiv \frac{1}{Z(v,\beta)} e^{-\beta U(y)+\gamma v \beta y}, \tag{4.34}$$

where $Z(v,\beta)$ is a normalization constant determined by

$$Z(v,\beta) = \int_{-L}^{L} dy e^{-\beta U(y)+\gamma v \beta y}. \tag{4.35}$$

The first result is about the convergence of the distribution function of y to this canonical distribution function in $L \to \infty$. More precisely, this statement is formulated as

$$p_{\text{st}}^{U,v,\beta}(y) \sim p_{\text{can}}^{U,v,\beta}(y) \tag{4.36}$$

for $y = YL$ with $Y \in \{Y : -1 < Y < 1\}$, where the definition of the symbol $\sim$ is the following: For given two functions $A(Y)$ and $B(Y)$, where both of them depend on L, we define $A(Y) \sim B(Y)$ for $-1 < Y < 1$ as $\lim_{L\to\infty}(1/L)\log[A(Y)/B(Y)] = 0$ for a given $Y \in \{Y : -1 < Y < 1\}$ fixed. We use this symbol $\sim$ throughout this chapter.

The second result is about the asymptotic form of the scaled cumulant generating function $G_{\text{scaled}}(h_w, h_q)$. This function always converges to a quadratic function in $L \to \infty$ limit,

$$\lim_{L\to\infty} G_{\text{scaled}}(h_w, h_q) = \gamma v^2 (h_w + h_q) + T\gamma v^2 (h_w + h_q)^2. \tag{4.37}$$

Here, we stress that this result is always true whenever the periodic potential satisfies the condition (4.7). We also mention that, from (4.23) and this relation,

$$\lim_{L\to\infty} \lim_{t\to\infty} G(h_w, h_q, t|p) = \gamma v^2 (h_w + h_q) + T\gamma v^2 (h_w + h_q)^2 \tag{4.38}$$

is satisfied for any h_w and h_q.

The system that we now consider was analyzed by Lebowitz and Spohn in Ref. [6]. They proved the fluctuation theorem of the work and the heat in this system. The theorem was written as

$$G_{\text{scaled}}(h_w, h_q) = G_{\text{scaled}}(-h_w, -\beta - h_q)|_{v\to -v}, \tag{4.39}$$

where $|_{v\to -v}$ means that the sign of v is reversed. We will re-derive (4.39) in Sect. 4.4 for the sake of completeness. From (4.39),

$$\lim_{L\to\infty}\lim_{t\to\infty} G(h_w, h_q, t|p) = \lim_{L\to\infty}\lim_{t\to\infty} G(-h_w, -\beta - h_q, t|p)|_{v\to -v} \tag{4.40}$$

is obtained. We note that this equality is consistent with the expression (4.38), which can be checked by a direct substitution.

From now on, we set the initial distribution function $p(y)$ to be the stationary distribution function $p_{\text{st}}^{U,v',\beta'}(y)$, where $\beta'(>0)$ and v' represent an inverse temperature and a velocity in a different system from the original one (β, v). The third result is about the asymptotic behaviour of the biased joint distribution function $P_{h_w,h_q}(y_0, y, t|p_{\text{st}}^{U,v',\beta'})$, when both of t and L are large. The initial part (about y_0) and the final part (about y) become independent, and the both parts take the canonical distribution function form with different temperatures and velocities:

$$P_{h_w,h_q}(y_0, y, t|p_{\text{st}}^{U,v',\beta'}) \sim p_{\text{can}}^{U,v_{\text{i}},\beta_{\text{i}}}(y_0)\,p_{\text{can}}^{U,v_{\text{f}},\beta_{\text{f}}}(y) + O(e^{-at}) \tag{4.41}$$

for $y_0 = Y_0 L$, $y = YL$ $(-1 < Y_0, Y < 1)$ with

$$\beta_{\text{i}} = \beta' - h_q, \tag{4.42}$$

$$v_{\text{i}} = \frac{v'\beta' - v(h_q + h_w)}{\beta' - h_q}, \tag{4.43}$$

$$\beta_{\text{f}} = \beta + h_q, \tag{4.44}$$

$$v_{\text{f}} = \frac{v(\beta + h_q + h_w)}{\beta + h_q}. \tag{4.45}$$

From these canonical distribution forms, we find that the inverse temperatures in the initial and the finial time can take negative values. Then below, we show that the excess cumulant generating function has different asymptotic behaviours in $L \to \infty$, depending on the sign of this inverse temperature. This is the fourth result: For large L, the excess cumulant generating function $H_{\text{ex}}(h_w, h_q|p_{\text{st}}^{U,v',\beta'})$ satisfies

$$H_{\text{ex}}(h_w, h_q|p_{\text{st}}^{U,v',\beta'}) = O(U(L)) \tag{4.46}$$

for $\beta_i < 0$ or $\beta_f < 0$, and

$$H_{\text{ex}}(h_w, h_q|p_{\text{st}}^{U,v',\beta'}) = O(1) \tag{4.47}$$

for $\beta_i > 0$ and $\beta_f > 0$. Therefore, from (4.23) and these relations, we find

$$\lim_{t\to\infty}\lim_{L\to\infty} G(h_w, h_q, t|p) = \infty \tag{4.48}$$

for $\beta_i < 0$ or $\beta_f < 0$, and

$$\lim_{t\to\infty}\lim_{L\to\infty} G(h_w, h_q, t|p) = \lim_{L\to\infty}\lim_{t\to\infty} G(h_w, h_q, t|p) \tag{4.49}$$

for $\beta_i > 0$ and $\beta_f > 0$. (4.49) means the following things: When β_i and β_f are both positive, the two limiting operations, which are $L \to \infty$ and $t \to \infty$, are interchangeable, and furthermore, the symmetry property of the fluctuation theorem is satisfied. However, on the other side, when β_i or β_f is negative, the two limiting operations become non-interchangeable. In other words, *if we take $t \to \infty$ first, the cumulant generating function satisfies the fluctuation theorem* (4.40). *On the other hand, if we take $L \to \infty$ first, the cumulant generating function diverges as shown in* (4.48). *This divergence corresponds to the van Zon–Cohen singularity* [10, 11].

4.3.1 Negative Inverse Temperature and the van Zon–Cohen Singularity

Here, we consider this negative inverse temperature in terms of the conditional distribution function that we introduced in Sect. 4.2.3.

First, we connect the condition q^* and the biasing field h_q, (4.29) and (4.30), in terms of the quadratic form of the cumulant generating function given as (4.37). The result is

$$h_q = \frac{q^* - \gamma v^2}{2T\gamma v^2}. \tag{4.50}$$

By using this relation, we connect the third result stated above, with the conditional distribution function (4.32), as

$$\frac{P(y_0, y, q, t|p_{\mathrm{st}}^{U,v'\beta'})}{P(q, t|p_{\mathrm{st}}^{U,v'\beta'})} \sim p_{\mathrm{can}}^{U,\tilde{v}_\mathrm{i},\tilde{\beta}_\mathrm{i}}(y_0)\, p_{\mathrm{can}}^{U,\tilde{v}_\mathrm{f},\tilde{\beta}_\mathrm{f}}(y) + \frac{1}{1/t} o(1/t) \tag{4.51}$$

for $y_0 = Y_0 L$, $y = YL$, where $Y, Y_0 \in \{Y, Y_0 : -1 < Y_0, Y < 1\}$, with the parameters defined as

$$\tilde{\beta}_\mathrm{i} = \beta' - \frac{q - \gamma v^2}{2T\gamma v^2}, \tag{4.52}$$

$$\tilde{v}_\mathrm{i} = \frac{v'\beta' + (q - \gamma v^2)/(2T\gamma v)}{\beta' - (q - \gamma v^2)/(2T\gamma v^2)}, \tag{4.53}$$

$$\tilde{\beta}_{\mathrm{f}} = \beta + \frac{q - \gamma v^2}{2T\gamma v^2}, \tag{4.54}$$

$$\tilde{v}_{\mathrm{f}} = v. \tag{4.55}$$

When we set q to be larger than $\gamma v^2(2\beta'/\beta+1)$, the initial inverse temperature (4.52) becomes negative. This means that, *with the condition of* $Q(t)/t > \gamma v^2(2\beta'/\beta+1)$, *the particle tends to climb down the potential at the initial time* 0. On the other hand, when we set q to be smaller than $-\gamma v^2$, the finial inverse temperature (4.54) becomes negative. This means that, *with the condition of* $Q(t)/t < -\gamma v^2$, *the particle tends to climb up the potential at the final time t.* In the following paragraph, we show the singularity of the cumulant generating function, from these rare trajectories, by using an intuitive argument.

Let us consider the situation that we measure the trajectories of the particle and evaluate $G(0, h_q, t|p_{\mathrm{st}}^{U,v',\beta'})$ from them. We set t to be sufficiently large. Then, the trajectories needed for the calculation of $G(0, h_q, t|p_{\mathrm{st}}^{U,v',\beta'})$ have to satisfy $Q(t)/t = q^*$ due to (4.30), where q^* is connected to h_q through (4.50). Now, we consider the case that β_{f} takes negative value. This indicates that the trajectories needed for the calculation of $G(0, h_q, t|p_{\mathrm{st}}^{U,v',\beta'})$ satisfy $y(t) = L$ in a high probability. Here, by using Jensen's inequality in $G(h_w, h_q, t|p_{\mathrm{st}}^{U,v',\beta'})$, we obtain

$$\begin{aligned} G(h_w, h_q, t|p_{\mathrm{st}}^{U,v',\beta'}) &\geq \frac{1}{t}\int_0^t dt \left[h_q \left\langle \dot{Q}(t)\right\rangle_{p_{\mathrm{st}}^{U,v',\beta'}} + h_w \left\langle \dot{W}(t)\right\rangle_{p_{\mathrm{st}}^{U,v',\beta'}}\right] \\ &= -\frac{h_q}{t}\left\langle U(y(t)) - U(y(0))\right\rangle_{p_{\mathrm{st}}^{U,v',\beta'}} + \frac{h_w + h_q}{t}\int_0^t dt \left\langle \dot{W}(t)\right\rangle_{p_{\mathrm{st}}^{U,v',\beta'}}, \end{aligned} \tag{4.56}$$

where (4.14) is used at the final line. In the right-hand side of this expression, we use the trajectories discussed above. With these trajectories (that satisfy $y(t) = L$), we approximate the first term of (4.56) as $-h_q U(L)/t$, where we used the fact that $U(y(0))$ is $O(1)$. We then neglect the second term as well by assuming that the particle moves around the bottom of the potential most of the time, then suddenly climbs up the potential before the time t and finally reaches the top of the potential at the time t. We therefore have

$$G(h_w, h_q, t|p) \gtrsim -\frac{h_q}{t}U(L), \tag{4.57}$$

which leads to (4.48).

For the case that β_{i} is negative, we can also discuss by following the same argument as the one for $\beta_{\mathrm{f}} < 0$. In this case, the difference is that the particle climbs down the potential from the top, instead of climbing up it from the bottom. From these arguments, we understand how hardly measurable trajectories can cause the van Zon–Cohen singularity.

4.4 Derivation of the Results

Here in this subsection, we derive the results stated above. In Sect. 4.4.1, we show how to analyze the system for a given finite L. Then, in Sect. 4.4.2, we perform a boundary layer analysis by considering the limit $L \to \infty$ in the framework that we developed in Sect. 4.4.1. Finally in Sect. 4.4.3, we derive the main results of this chapter by using the result of the boundary layer analysis.

4.4.1 The Method of the Largest Eigenvalue Problem and the Cole–Hopf Transformation

First, we define an operator $\mathcal{L}^{(y)}_{h_w,h_q}$ as

$$\begin{aligned}\mathcal{L}^{(y)}_{h_w,h_q} \cdot \varphi = &-\frac{\partial}{\partial y}\left[\left(-\frac{1}{\gamma}\frac{\partial}{\partial y}U(y) + v\right)\varphi\right] + h_w v\left(\frac{\partial}{\partial y}U(y)\right)\varphi \\ &+ h_q\left[\frac{1}{\gamma}\left(\frac{\partial U(y)}{\partial y}\right)^2 - \frac{T}{\gamma}\frac{\partial^2}{\partial y^2}U(y)\right]\varphi + \frac{T}{\gamma}\frac{\partial^2}{\partial y^2}\varphi \\ &+ \frac{T}{\gamma}\left(\frac{\partial U(y)}{\partial y}\right)^2 (h_q)^2\varphi + \frac{2T}{\gamma}h_q\frac{\partial}{\partial y}\left[\frac{\partial U(y)}{\partial y}\varphi\right]. \end{aligned} \tag{4.58}$$

We then denote the eigenfunctions of this operator by ψ_n $(n = 0, 1, 2, \ldots)$ and the corresponding eigenvalues to them by μ_n $(n = 0, 1, 2, \ldots)$. Here we note that the eigenvalues are labeled with a descending order with the label n ascending: $\mathrm{Re}(\mu_n) \le \mathrm{Re}(\mu_m)$ for $n > m$, where $\mathrm{Re}(a)$ is the real part of a. We also consider the adjoint operator of $\mathcal{L}^{(y)}_{h_w,h_q}$ given as

$$\begin{aligned}\mathcal{L}^{(y)\dagger}_{h_w,h_q} \cdot \varphi = &\left(-\frac{1}{\gamma}\frac{\partial}{\partial y}U(y) + v\right)\frac{\partial}{\partial y}\varphi + h_w v\left(\frac{\partial}{\partial y}U(y)\right)\varphi \\ &+ h_q\left[\frac{1}{\gamma}\left(\frac{\partial U(y)}{\partial y}\right)^2 - \frac{T}{\gamma}\frac{\partial^2}{\partial y^2}U(y)\right]\varphi + \frac{T}{\gamma}\frac{\partial^2}{\partial y^2}\varphi \\ &+ \frac{T}{\gamma}\left(\frac{\partial U(y)}{\partial y}\right)^2 (h_q)^2\varphi - \frac{2T}{\gamma}h_q\left(\frac{\partial U(y)}{\partial y}\right)\frac{\partial}{\partial y}\varphi. \end{aligned} \tag{4.59}$$

Again, the eigenfunctions of $\mathcal{L}^{(y)\dagger}_{h_w,h_q}$ are denoted by ϕ_n $(n = 0, 1, 2, \ldots)$ and the corresponding eigenvalues to them are denoted by ν_n $(n = 0, 1, 2, \ldots)$. Here, we can set

$$\nu_n = (\mu_n)^* \tag{4.60}$$

without loss of generality. By using the Perron-Frobenius theory, following theorems are satisfied: The largest eigenvalues of $\mathcal{L}^{(y)\dagger}_{h_w,h_q}$ and $\mathcal{L}^{(y)}_{h_w,h_q}$ are real and do not degenerate, which indicates

$$\nu_0 = \mu_0. \tag{4.61}$$

Furthermore, the eigenfunctions corresponding to the largest eigenvalue are real. See the Appendix B of Ref. [30] for the derivation. The orthonormal conditions for these eigenfunctions are written as

$$\int_{-L}^{L} dy(\phi_n(y))^* \psi_m(y) = \delta_{n,m} \tag{4.62}$$

$(n, m = 0, 1, 2, \ldots)$, where $\delta_{n,m}$ is the Kronecker δ.

Now, we define $q_{h_w,h_q}(y_0, y, t|p)$ as

$$q_{h_w,h_q}(y_0, y, t|p) = e^{tG(h_w,h_q,t|p)} P_{h_w,h_q}(y_0, y, t|p). \tag{4.63}$$

As shown in Appendix C.1, $\mathcal{L}^{(y)}_{h_w,h_q}$ is the time evolution operator of $q_{h_w,h_q}(y_0, y, t|p)$, which means

$$\frac{\partial}{\partial t} q_{h_w,h_q}(y_0, y, t|p) = \mathcal{L}^{(y)}_{h_w,h_q} \cdot q_{h_w,h_q}(y_0, y, t|p). \tag{4.64}$$

Then, we expand $q_{h_w,h_q}(y_0, y, t|p)$ in terms of the eigenfunctions $(\psi_n(y))_{n=0}^{\infty}$ and substitute it into this time evolution equation. By solving it, we determine the time dependence of the expansion coefficients in $q_{h_w,h_q}(y_0, y, t|p)$ with an initial condition $q_{h_w,h_q}(y_0, y, 0|p) = p(y_0)\delta(y - y_0)$. The result is

$$q_{h_w,h_q}(y_0, y, t|p) = p(y_0) \sum_{n=0}^{\infty} (\phi_n(y_0))^* \psi_n(y) e^{\mu_n t}. \tag{4.65}$$

Here, we consider the behaviour of this function when t is large. Since the exponential term $e^{\mu_n t}$ in (4.65) determines the time-dependency of $q_{h_w,h_q}(y_0, y, t|p)$, we find that the $n = 0$ term (largest eigenvalue term) gives a dominant contribution. Thus, with (4.63), we obtain

$$P_{h_w,h_q}(y_0, y, t|p) = e^{(\mu_0 - G(h_w,h_q,t|p))t} \left[p(y_0)\phi_0(y_0)\psi_0(y) + O\left(e^{-(\mu_0 - \mathrm{Re}(\mu_1))t}\right)\right]. \tag{4.66}$$

This is the asymptotic form of the biased distribution. Then, the expression of the cumulant generating function is derived from the integration of this asymptotic form with respect to y_0 and y,

$$G(h_w, h_q, t|p) = \mu_0 + \frac{1}{t} \log c_0 \tilde{c}_0 + O\left(e^{-(\mu_0 - \mathrm{Re}(\mu_1))t}\right), \tag{4.67}$$

where c_0 and $\tilde{c}_0$ are defined as

$$c_0 = \int_{-L}^{L} dy \phi_0(y) p(y), \tag{4.68}$$

$$\tilde{c}_0 = \int_{-L}^{L} dy \psi_0(y). \tag{4.69}$$

Finally, by comparing (4.23) with (4.67), we arrive at the expression of the scaled and excess cumulant generating functions,

$$G_{\text{scaled}}(h_w, h_q) = \mu_0, \tag{4.70}$$

$$H_{\text{ex}}(h_w, h_q|p) = \log c_0 \tilde{c}_0. \tag{4.71}$$

Here, (4.70) is well-known. See [27, 29] for example. (4.70) has been used in many applications, such as the analysis of the fluctuation theorem [6], for example. Also the result (4.71) was used for the calculation of an excess heat in Ref. [28] recently. We note that the expression (4.66) leads to an expression of the biased joint distribution function, such as

$$q_{h_w,h_q}(y_0, y, t|p) \sim p(y_0)\phi_0(y_0)\psi_0(y) e^{-H_{\text{ex}}(h_w,h_q|p)} \tag{4.72}$$

when t is large, due to (4.70) and (4.71). We also note that essentially the same result was discussed in Ref. [31].

Next, by using the specific property of Langevin equation, we write down the formula to obtain ϕ_0 and ψ_0 appeared above. Especially, we apply the Cole-Hopf transformation to the largest eigenvalue problems $\mathcal{L}^{(y)}_{h_w,h_q}$ and $\mathcal{L}^{(y)\dagger}_{h_w,h_q}$, then we rewrite them as a non-linear eigenvalue problem. The details of the derivation is developed in Appendix C.2. Here we show only the results. We define a non-linear operator $\mathcal{M}_{h,v}$ as

$$\mathcal{M}_{h,v} \cdot \varphi = 2Thv \frac{\partial U}{\partial y} + \frac{1}{2\gamma}\varphi^2 + \left(-\frac{1}{\gamma}\frac{\partial U}{\partial y} + v \right) \varphi + \frac{T}{\gamma}\frac{\partial}{\partial y}\varphi. \tag{4.73}$$

Then, with this operator, we define a non-linear eigenvalue problem

$$\mathcal{M}_{h,v} \cdot w_{h,v} = K_{h,v}, \tag{4.74}$$

where the constant $K_{h,v}$ and the periodic function (eigenfunction) $w_{h,v}(y)$ are simultaneously determined from the boundary condition $w_{h,v}(-L) = w_{h,v}(L)$ and the normalisation condition

$$\int_{-L}^{L} dy w_{h,v}(y) = 0. \tag{4.75}$$

From this eigenvalue problem, we define a function $w_{h,v}$. This is connected to ϕ_0 and ψ_0 as follows: First, we introduce a potential function of $w_{h,v}(y)$ by

$$V_{h,v}(y) = -\int_0^y dz w_{h,v}(z) + \text{const.} \tag{4.76}$$

Then, we can derive the following relations

$$\phi_0(y) = \frac{1}{C} \exp\left[h_q U(y) - \frac{V_{h_q+h_w,v}(y)}{2T}\right], \tag{4.77}$$

$$\psi_0(y) = \frac{1}{\tilde{C}} \exp\left[-\left(h_q + \beta\right) U(y) - \frac{V_{-\beta-h_q-h_w,-v}(y)}{2T}\right], \tag{4.78}$$

$$G_{\text{scaled}}(h_w, h_q) = \frac{K_{h_q+h_w,v}}{2T}, \tag{4.79}$$

where the coefficients $(C)^* \tilde{C}$ are the normalisation constant (4.62). See Appendix C.2 for the details of the derivation of these relations. We mention that similar arguments were presented in Refs. [30, 32].

4.4.2 Boundary Layer Analysis in Large L Limit

Here, we evaluate the asymptotic behaviour of $w_{h,v}(y)$ and $K_{h,v}$ when L is large. We especially use the boundary layer analysis technique [33].[1]

First, we define scaled $w_{h,v}$ function, $\tilde{w}_{h,v}(Y)$, as

$$\tilde{w}_{h,v}(Y) \equiv w_{h,v}(LY), \tag{4.80}$$

where $-1 \leq Y \leq 1$. With this scaled function, we rewrite the left-hand side of (4.74) as

$$2Thv\frac{\partial U(YL)}{\partial y} + \frac{1}{2\gamma}(\tilde{w}_{h,v})^2 + \left(-\frac{1}{\gamma}\frac{\partial U(YL)}{\partial y} + v\right)\tilde{w}_{h,v} + \frac{1}{L}\frac{T}{\gamma}\frac{\partial}{\partial Y}\tilde{w}_{h,v}. \tag{4.81}$$

Now, we regard L^{-1} as a perturbation parameter. In this equation, we find that the coefficient of $\partial\tilde{w}_{h,v}/\partial Y$ is $O(L^{-1})$. We thus expect that there is a domain of Y, I_b, where $\partial\tilde{w}_{h,v}/\partial Y$ becomes large, (or in other words, $\tilde{w}_{h,v}(Y)$ changes rapidly), due to the reason that the periodic boundary condition and the normalisation condition of this nonlinear eigenvalue problem have to be satisfied. This is written as

[1] In this calculation, we use the condition (4.7).

$$\left|\frac{\partial \tilde{w}_{h,v}(Y)}{\partial Y}\right| \gg \left|\tilde{w}_{h,v}(Y)\right| \tag{4.82}$$

for $Y \in I_{\mathrm{b}}$. It is a convention to call this region a boundary layer, when the width of the region I_{b} becomes 0 in $L \to \infty$ limit. We assume the existence of a boundary layer in this problem, and use a well-known basic strategy to solve this problem. It is, (i) constructing the solution inside the boundary layer that is called inner solution, (ii) constructing the solution outside the boundary layer that is called outer solution, and (iii) matching those solutions asymptotically with keeping the continuity and the boundary conditions satisfied. See Ref. [33] for the details of this boundary layer analysis. In this chapter, we consider only the outer solution and obtain the leading order of $\tilde{w}_{h,v}(Y)$ by using a few assumptions. We show the results here. The derivation is developed in Appendix C.3.

By using the condition (4.7), we obtain an asymptotic form in $L \to \infty$ as

$$\tilde{w}_{h,v}(Y) = \begin{cases} 2T\gamma hv & -1 \le Y \le a_- \\ 2\gamma\left[-v(1+Th) + (1/\gamma)\partial U(YL)/\partial y\right] & a_- \le Y \le 1 \end{cases} \tag{4.83}$$

for $hv \le 0$ and

$$\tilde{w}_{h,v}(Y) = \begin{cases} 2\gamma\left[-v(1+Th) + (1/\gamma)\partial U(YL)/\partial y\right] & -1 \le Y \le a_+ \\ 2T\gamma hv & a_+ \le Y \le 1 \end{cases} \tag{4.84}$$

for $hv \ge 0$, where the coefficients a_+ and a_- are determined from the normalisation condition (4.75):

$$\gamma v a_-\left(1 + 2Th\right) - \gamma v + \frac{U(L) - U(a_- L)}{L} = 0 \tag{4.85}$$

and

$$\gamma v a_+\left(1 + 2Th\right) + \gamma v - \frac{U(La_+) - U(-L)}{L} = 0. \tag{4.86}$$

We plot these functions for the case of U_{harmo} and U_{quart} in Fig. 4.2 as green dashed lines and purple dotted lines. In the same figure, we also plot the numerical results of $\tilde{w}_{h,v}(Y)$ obtained from the method used in Ref. [30]. These lines are plotted as red solid lines. The figure shows the coincidence between these analytical and numerical results.

Next, by substituting an expression of $\tilde{w}_{h,v}(Y) = 2T\gamma hv$ into the eigenvalue Eq. (4.74), we find that $K_{h,v}$ is

$$K_{h,v} = 2T\left(\gamma v^2 h + T\gamma v^2 h^2\right), \tag{4.87}$$

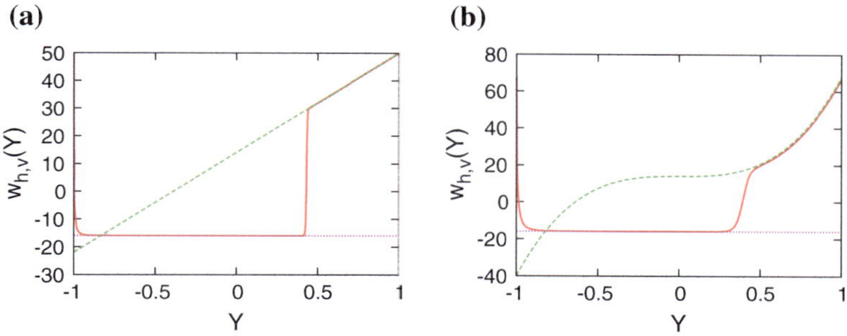

Fig. 4.2 Numerical and analytical results of $\tilde{w}_{h,v}(Y)$ for a harmonic potential $U_{\text{harmo}}(y) = (k/2)y^2$ (**a**) and for a quartic potentials $U_{\text{quart}}(y) = (k_4/4)y^4$ (**b**). Quantities are converted to dimensionless forms with setting $\gamma = T = v = 1$. We set $k = 1$, $k_4 = 1$ and $h = -8$, and also, we set $L = 18$ for $U_{\text{harmo}}(y)$ and $L = 3$ for $U_{\text{quart}}(y)$. The *green dashed lines* and *purple dotted lines* corresponds to the analytical results (4.83). These lines are $14 + 36Y$ and -16 for $U_{\text{harmo}}(y)$, and $14 + 54Y^3$ and -16 for $U_{\text{quart}}(y)$. The *red solid lines* are numerical results. Reprinted with permission from Ref. [20]. Copyright 2012 by American Physical Society

which leads to

$$G_{\text{scaled}}(h_w, h_q) = \gamma v^2(h_w + h_q) + T\gamma v^2(h_w + h_q)^2. \tag{4.88}$$

We check this result in Fig. 4.3, by comparing it with numerical ones. We stress that (4.88) is always satisfied for any periodic potentials, as long as the potential satisfies the condition (4.7).

4.4.3 Derivation of the Main Results

Finally, we derive the main results of this chapter, which was explained in Sect. 4.3.

First, from (4.85) and (4.86), we find

$$\lim_{L\to\infty} a_\pm = \mp 1, \tag{4.89}$$

which indicates

$$\lim_{L\to\infty} \tilde{w}_{h,v}(Y) = 2T\gamma hv \tag{4.90}$$

for fixed Y ($-1 < Y < 1$). From this relation, we evaluate the limiting value of $V_{h_q+h_w,v}(y)$ and $V_{-\beta-h_q-h_w,-v}(y)$ in $L \to \infty$, as

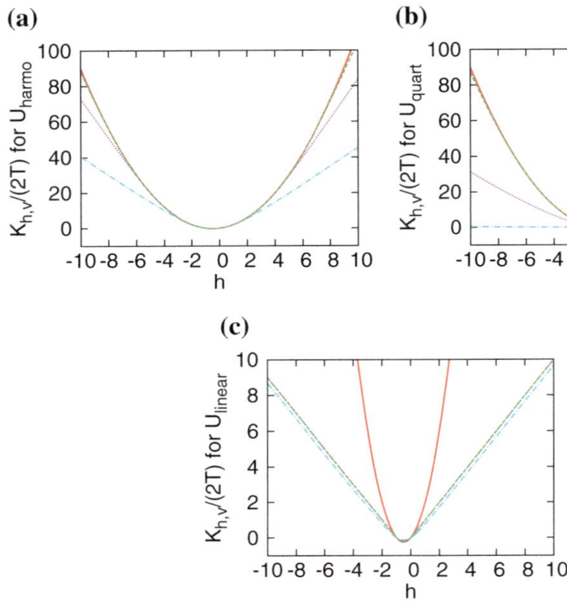

Fig. 4.3 The eigenvalue $K_{h,v}/(2T)$ for a harmonic potential $U_{\text{harmo}}(y) = (k/2)y^2$ (**a**), a quartic potential $U_{\text{quart}}(y) = (k_4/4)y^4$ (**b**), and a linear potential $U_{\text{linear}}(y) = k_1|y|$ (**c**). Quantities are converted to dimensionless forms by setting $\gamma = T = v = 1$. We set $k = 1$, $k_4 = 1$ and $k_1 = 1$, and $K_{h,v}/(2T)$ is numerically evaluated. For U_{harmo}, *dashed dotted line* (*aqua blue*), *dotted line* (*purple*) and *dashed line* (*green*) correspond to $L = 6$, $L = 12$ and $L = 18$, respectively. For U_{quart}, these lines correspond to $L = 6$, $L = 12$ and $L = 18$, and for U_{linear}, these lines correspond to $L = 5$, $L = 25$ and $L = 50$. In all figure, we also plot *red solid lines*. This corresponds to $h + h^2$, which is an analytical result predicted by (4.87). We can see that $K_{h,v}$ approaches $h+h^2$ for U_{harmo} and U_{quart}, as L becomes larger. However, this convergence does not occur for U_{linear}. This is due to the condition (4.7): U_{harmo} and U_{quart} satisfy it, but U_{linear} does not. Reprinted with permission from Ref. [20]. Copyright 2012 by American Physical Society

$$\begin{aligned}\lim_{L\to\infty} V_{h_q+h_w,v}(YL)/L &= -\lim_{L\to\infty}\frac{1}{L}\left[\int_0^{YL} dz w_{h_q+h_w,v}(z) + \text{const.}\right] \\ &= -2T\gamma v(h_q + h_w)Y + \text{const.}\end{aligned} \tag{4.91}$$

for fixed Y $(-1 < Y < 1)$ and

$$\begin{aligned}\lim_{L\to\infty} V_{-\beta-h_q-h_w,-v}(y)/L &= -\lim_{L\to\infty}\frac{1}{L}\left[\int_0^{YL} dz w_{-\beta-h_q-h_w,-v}(z) + \text{const.}\right] \\ &= -2T\gamma v(\beta + h_q + h_w)Y + \text{const.}\end{aligned} \tag{4.92}$$

for fixed Y $(-1 < Y < 1)$. We thus obtain $p(y_0)\phi_0(y_0)$ and $\psi_0(y)$ from these expressions by combining them with (4.77) and (4.78). The results are

$$p(y_0)\phi_0(y_0) \sim \frac{1}{C'} \exp\left[h_q U(y_0) + \log p(y_0) + \gamma v \left(h_q + h_w\right) y_0\right] \tag{4.93}$$

for $y_0 = Y_0 L$ $(-1 < Y_0 < 1)$ and

$$\psi_0(y) \sim \frac{1}{\tilde{C}'} \exp\left[-\left(\beta + h_q\right) U(y) + \gamma v(\beta + h_q + h_w) y\right] \tag{4.94}$$

for $y = YL$ $(-1 < Y < 1)$, where C' and $\tilde{C}'$ are normalisation constants determined from (4.62).

Now, we derive the main results in this chapter. First we derive (4.36). Since this is the result about the unbiased system ($h_w = h_q = 0$), we set $h_w = h_q = 0$ in (4.94). Then, we use the obtained expression of the biased distribution function (4.66), which leads to (4.36). This means that the stationary distribution function $p_{\mathrm{st}}^{U,v,\beta}(y)$ in $L \to \infty$ is the canonical distribution function.

Then, we notice that the asymptotic form of the scaled cumulant generating function (4.37), which is the second result, is nothing but (4.88). Furthermore, by looking at the expression of $\phi_0(y)$ and $\psi_0(y)$ in terms of the potential function $V_{h,v}(y)$, which are (4.77) and (4.78), we find

$$\phi_0(y) = \psi_0(y)|_{(h_w, h_q, v) \to (-h_w, -\beta - h_q, -v)}. \tag{4.95}$$

From this relation with (4.61) and (4.70), the fluctuation theorem for the scaled cumulant generating function is obtained as

$$G_{\mathrm{scaled}}(h_w, h_q) = G_{\mathrm{scaled}}(-h_w, -\beta - h_q)|_{v \to -v}, \tag{4.96}$$

which is (4.39).

Next, we set the initial distribution function to be the stationary distribution function $p_{\mathrm{st}}^{U,v',\beta'}(y)$. Then, we obtain (4.41), which is the third result, by substituting (4.93) and (4.94) into (4.66).

Finally, we derive (4.46) and (4.47). We substitute the expression of $\phi_0(y)$ and $\psi_0(y)$, which are (4.77) and (4.78), into (4.71). Then, by noticing the normalization condition (4.62), we obtain

$$
\begin{aligned}
H_{\mathrm{ex}}(h_w, h_q|p) = \log \int_{-L}^{L} dy \exp\left[h_q U(y) + \log p(y) - \frac{V_{h_q+h_w,v}(y)}{2T}\right] \\
+ \log \int_{-L}^{L} dy \exp\left[-\left(h_q+\beta\right)U(y) - \frac{V_{-\beta-h_q-h_w,-v}(y)}{2T}\right] \\
- \log \int_{-L}^{L} dy \exp\left[-\beta U(y) - \frac{1}{2T}\left(V_{h_q+h_w,v}(y) + V_{-\beta-h_q-h_w,-v}(y)\right)\right].
\end{aligned}
\tag{4.97}
$$

We substitute the potential functions $V_{h,v}(y)$ appearing in these expressions, by the asymptotic forms shown in (4.91) and (4.92). After this substitution, by defining

$$
H \equiv \log \int_{-L}^{L} dy \exp\left[h_q U(y) + \log p(y) + \gamma v \left(h_q + h_w\right) y\right], \tag{4.98}
$$

$$
\tilde{H} \equiv \log \int_{-L}^{L} dy \exp\left[-\left(h_q+\beta\right)U(y) + \gamma v(\beta + h_q + h_w) y\right] \tag{4.99}
$$

and

$$
Z \equiv -\log \int_{-L}^{L} dy \exp\left[-\beta U(y) + \gamma v(\beta + 2h_q + 2h_w) y\right], \tag{4.100}
$$

we write down the excess cumulant generating function as

$$
H_{\mathrm{ex}}(h_w, h_q|p) \simeq H + \tilde{H} + Z. \tag{4.101}
$$

Now we consider the dependence of L in each terms of this expression. First, Z is always $O(1)$:

$$
Z = O(1). \tag{4.102}
$$

The other terms, H and $\tilde{H}$, depend on the parameter h_q. $\tilde{H}$ depends on the sign of $h_q + \beta$. It becomes $O(1)$ for $h_q + \beta > 0$, but $O(U(L))$ for $h_q + \beta < 0$,

$$
\tilde{H} = \begin{cases} \mathcal{O}(1) & h_q > -\beta \\ O(U(L)) & h_q < -\beta. \end{cases} \tag{4.103}
$$

H depends on the initial distribution function $p(y)$ as well. For seeing this dependence, we define a parameter β_p as

$$
\beta_p \equiv \lim_{L\to\infty} \frac{-\log p(L)}{U(L)}, \tag{4.104}
$$

which takes a value from 0 to ∞. This parameter represents an effective temperature of the initial distribution function. By using this effective temperature β_p, we obtain

$$H = \begin{cases} O(U(L)) & h_q > \beta_p \\ \mathcal{O}(1) & h_q < \beta_p. \end{cases} \tag{4.105}$$

Then, from these L-dependencies (4.102), (4.103) and (4.105), we arrive at the dependence of L for the excess cumulant generating function $H_{\text{ex}}(h_w, h_q|p)$,

$$H_{\text{ex}}(h_w, h_q|p) = \begin{cases} O(U(L)) & h_q < -\beta \\ O(1) & -\beta < h_q < \beta_p \\ O(U(L)) & h_q > \beta_p. \end{cases} \tag{4.106}$$

In this expression, by setting $p = p_{\text{st}}^{U,v',\beta'}(y)$, we obtain (4.46) and (4.47).

4.5 Conclusions

In this chapter, the fluctuation of the work and the heat of a Brownian particle on a moving periodic potential was studied. Especially the special limit, where the particle was trapped by the periodic potential, was considered in order to study the van Zon–Cohen singularity for an extended fluctuation theorem. As a result, we discovered that the conditional distribution function, given the heat dissipation rate $Q(t)/t = q$, became a canonical distribution function. Also, we discovered that, when q was larger (or smaller) than a special value, the inverse temperature of the conditional distribution function became negative. This meant that the particle tended to climb up (climb down) the potential, where this tendency led to the van Zon–Cohen singularity.

Before ending this chapter, we briefly discuss the possibility of observing the singularity, taking place not only in the heat but also in other quantities. We found that the singularity occurred due to the non-interchangeability of two types of limits. The first one was the large observation time limit, and the second one was the trapped particle limit. We showed that these two limits became non-interchangeable, which leads to the van Zon–Cohen singularity. This indicates that we may find another system by looking for the one that has the same limit-structure as the one studied in this chapter. In other words, we look for the system re-defined in a limit from another system, where this limit is non-interchangeable with the large observation time limit, in order to find another example of van Zon–Cohen singularity. We would like to explore such a system for a deeper understanding of fluctuation in nonequilibrium physics.

References

1. D.J. Evans, E.G.D. Cohen, G.P. Morriss, Phys. Rev. Lett. **71**, 2401 (1993)
2. G. Gallavotti, E.G.D. Cohen, Phys. Rev. Lett. **74**, 2694 (1995)
3. J. Kurchan, J. Phys. A **31**, 3719 (1998)
4. C. Maes, J. Stat. Phys. **95**, 367 (1999)
5. G.E. Crooks, Phys. Rev. E **60**, 2721 (1999)
6. J.L. Lebowitz, H. Spohn, J. Stat. Phys. **95**, 333 (1999)
7. K. Sekimoto, Lecture Notes in Physics, *Stochastic Energetics*, vol. 799 (Springer, Berlin, 2010)
8. G.M. Wang, E.M. Sevick, E. Mittag, D.J. Searles, D.J. Evans, Phys. Rev. Lett. **89**, 050601 (2002)
9. R. van Zon, E.G.D. Cohen, Phys. Rev. E **67**, 046102 (2003)
10. R. van Zon, E.G.D. Cohen, Phys. Rev. Lett. **91**, 110601 (2003)
11. R. van Zon, E.G.D. Cohen, Phys. Rev. E **69**, 056121 (2004)
12. N. Garnier, S. Ciliberto, Phys. Rev. E **71**, 060101(R) (2005)
13. F. Bonetto, G. Gallavotti, A. Giuliani, F. Zamponi, J. Stat. Phys. **123**, 39 (2006)
14. M. Baiesi, T. Jacobs, C. Maes, N.S. Skantzos, Phys. Rev. E **74**, 021111 (2006)
15. P. Visco, J. Stat. Mech. P06006 (2006)
16. R.J. Harris, A. Rákos, G.M. Schütz, Europhys. Lett. **75**, 227 (2006)
17. A. Rákos, R.J. Harris, J. Stat. Mech. P05005 (2008)
18. A. Puglisi, L. Rondoni, A. Vulpiani, J. Stat. Mech. P08010 (2006)
19. J.D. Noh, J.-M. Park, Phys. Rev. Lett. **108**, 240603 (2012)
20. T. Nemoto, Phys. Rev. E **85**, 061124 (2012)
21. L.P. Faucheux, G. Stolovitzky, A. Libchaber, Phys. Rev. E **51**, 5239 (1995)
22. V. Blickle, T. Speck, C. Lutz, U. Seifert, C. Bechinger, Phys. Rev. Lett. **98**, 210601 (2007)
23. J.R. Gomez-Solano, A. Petrosyan, S. Ciliberto, R. Chetrite, K. Gawędzki, Phys. Rev. Lett. **103**, 040601 (2009)
24. S. Toyabe, T. Sagawa, M. Ueda, E. Muneyuki, M. Sano, Nat. Phys. **6**, 988 (2010)
25. C.W. Gardiner, *Handbook of Stochastic Methods for Physics, Chemistry, and the Natural Sciences* (Springer, Berlin, 1983)
26. D. Ruelle, *Thermodynamic Formalism* (Addison-Wesley, Reading, 1978)
27. H. Touchette, Phys. Rep. **478**, 1 (2009)
28. T. Sagawa, H. Hayakawa, Phys. Rev. E **84**, 051110 (2011)
29. A. Dembo, O. Zeitouni, *Large Deviations Techniques and Applications* (Springer, New York, 1998)
30. T. Nemoto, S. Sasa, Phys. Rev. E **84**, 061113 (2011)
31. R.L. Jack, P. Sollich, Prog. Theor. Phys. Suppl. **184**, 304 (2010)
32. T. Nemoto, S. Sasa, Phys. Rev. E **83**, 030105(R) (2011)
33. C.M. Bender, S.A. Orszag, *Advanced Mathematical Methods for Scientists and Engineers* (McGraw-Hill, New York, 1978)

Chapter 5
Conclusions and Future Perspectives

In this thesis, we have studied *phenomenological structure for the large deviation principle* in time-series statistics. For equilibrium statistical mechanics, an exponentially biased ensemble was directly connected to the distribution function in another equilibrium system. We called this structure the phenomenological structure for the large deviation principle, and then we sought for another example of large deviation statistics that also possesses this structure. We focused on time-series statistics and set three problems. One was about a *rare event sampling method* discussed in Chap. 2. The second one was about dynamical phase transitions in KCMs for understanding glassy features, in Chap. 3. The last one was about the application of this structure to van Zon-Cohen extended fluctuation theorem, in Chap. 4. For all of these works, we discussed several future problems, in the conclusion of each chapter. Here in the final conclusion, beyond each subject, we discuss general possibility of large deviation statistics in time-series statistics.

The fluctuation theorem found in 1993 was a turning point in non-equilibrium physics [1]. One of the astonishing point was that the theorem could be regarded as an extension of several important results in statistical physics: The second law of thermodynamics is well-known as a basis of thermodynamics, which gives a bound to the expected value of entropy production $\langle \Delta S \rangle$ in total systems, such as $\langle \Delta S \rangle > 0$. The fluctuation theorem states that there is an equality behind this inequality. That equality includes the higher order moments of ΔS, and written as a simple form as $\left\langle e^{-\Delta S} \right\rangle = 1$. (The second law of thermodynamics directly follows this equality.) Furthermore, by considering liner response regime in a non-equilibrium condition, one can derive the liner response formula [2] from this theorem. As expected from the form of this theorem, this relation is expressed as a formula of large deviation functions [3]. Due to the finding of this fluctuation theorem, formal studies in non-equilibrium physics have been enhanced. Also, related to this fluctuation theorem, some exact solutions in mathematical models were found. These findings are important, but there is a tendency that theory proceeds real physics. For breaking this trend,

T. Nemoto, *Phenomenological Structure for the Large Deviation Principle in Time-Series Statistics*, Springer Theses,
DOI 10.1007/978-981-287-811-3_5

it is a good opportunity now to apply these obtained results and techniques to interesting physical systems, where some rare fluctuations play an important role. Below, as an example of it, we introduce a fully developed turbulence.

The studies of fully developed turbulence are old but still there are several interesting questions remained [4]. In the study using numerical simulations such as solving Navier–Stokes equation, we need to give up to follow each realisation of this equation for a detailed prediction, because this dynamical equation has chaoticity. But instead of the detail of each realisation, we can discuss statistical properties originated from the randomness of this chaoticity. There were several studies that focus on the difference between this randomness with the randomness originated from equilibrium thermal reservoirs. The examples of the comparisons are the diffusion coefficient of the Brownian particle immersed in the turbulent flow and thermal reservoirs, entropy production of these particles to ask if the fluctuation theorem holds in fully developed turbulence, and so on. One of the famous statistical properties of fully developed turbulence is Kolmogorov law, which was found in 1941 [4]. In homogeneous isotropic fully developed turbulence, the velocity difference between two points is denoted by $dv(r)$, where r is the distance between them. Then, Kolmogorov derived a scaling law with an assumption that the turbulence is scale invariant. The relation is written as $\langle dv(r)^n \rangle \sim Er^{n/3}$, where E is the energy injection to the system. This law was confirmed in both of real experiments and numerical simulations for small n. However, it was also observed that there was a deviation from this law for large n (higher order moments) [4]. This deviation is believed to be related to the intermittency of the turbulence. The main obstacle here is that the observation of higher order moments becomes harder as n becomes larger.

Even though it is believed that the deviation of Kolmogorov law is related to the higher order moments, namely, rare fluctuations, there has been no research to connect this deviation and the large deviation property in time-series statistics. (with the large deviation property in with scale variables, we say "yes". See Ref. [4] for example.) This is mainly due to the fact that it seems difficult, at first glance, to find the connection between the large deviation property (the rare fluctuation of *time-averaged quantity*) and instantaneous higher order fluctuation. However, it is possibly true that the intermittency is related to rare trajectories somehow, because the assembly of those intermittent trajectories make one long-time trajectory, which is characterized by the large deviation principle of time-averaged quantity. If one succeeds to make a map from the problem to the one in large deviation statistics, there is an important benefit from it. That is, one can utilize several techniques developed in non-equilibrium physics to them.

As the beginning of this research, we start with a simple toy model that shows the Kolmogorov law and its deviation. As an example of such a model, we consider the shell model [5]. This is an effective dynamical model that describes the Fourier components of the velocity fields in turbulence. With introducing a scaled wave number and the corresponding variables (shell variables), the number of the variables are incredibly reduced compared with the one in the direct simulation of Navier–Stokes equation. Indeed, around 20 variables, we can observe the Kolmogorov law for small n, and the deviation of Kolmogorov law for large n, which appears in the same way as the one observed in real turbulence (observed in Navier–Stokes equation or real experiments) [5]. To this system, we apply the population dynamics method to calculate the large deviation function, which was proposed by Giardina et al. in [6, 7]. We will calculate several large deviation functions, and investigate which time-averaged quantity is related to the deviation of Kolmogorov law. The examples of the time-averaged quantities may be time-averaged energy, time-averaged energy current between two adjacent scales, the largest Lyapunov exponent, and so on. Since the intermittency is related to the deviation of Kolmogorov law, by controlling the intermittency with the population dynamics, it is certainly possible that we will find the connection between the deviation of Kolmogorov law and the large deviation properties of time-averaged quantities.

After we understand the relation between large deviation functions and the deviation of Kolmogorov law, we next apply our phenomenological method developed in Chap. 2 to this turbulent system. We seek for the corresponding steady state to the biased ensemble that shows, on which the deviation of Kolmogorov law is controlled. Since the proof of our method was done on the system described by Markov dynamics, the application is not straightforward. Also, we expect that we need to find an effective description introduced in Chap. 2. But, if we succeed in the application of our method and if the effective description is simple enough to be implemented in real experiments, we can propose to experimentalists the method to observe the deviation of Kolmogorov law in operational manner, meaning that, in the way that we change the settings of the experiment, the deviation of Kolmogorov law is observed as a typical property of its new stationary state.

This is one of the challenging goals of our research: Rare events are hardly observed, but they are important in physics. Due to the lack of the method to accelerate the observation of rare evens, approaching these problems with experiments is difficult. The rare event sampling method for real experiments proposed in Chap. 2 is applied to them and such experimental approaches take place. After these successes, further studies follow. Finally, a new field *called rare-event samplings in real experiments* appears. We believe that this thesis might be the first step of such an attempt.

References

1. D.J. Evans, E.G.D. Cohen, G.P. Morriss, Phys. Rev. Lett. **71**, 2401 (1993)
2. R. Kubo, M. Toda, N. Hashitsume, *Statistical Physics II: Nonequilibrium Statistical Mechanics* (Springer, Berlin, 1991)
3. J.L. Lebowitz, H. Spohn, J. Stat. Phys. **95**, 333 (1999)
4. U. Frisch, *Turbulence: The Legacy of A. N. Kolmogorov* (Cambridge University, Cambridge, 1995)
5. L. Biferale, Annu. Rev. Fluid Mech. **35**, 441 (2003)
6. C. Giardina, J. Kurchan, L. Peliti, Phys. Rev. Lett. **96**, 120603 (2006)
7. C. Giardina, J. Kurchan, V. Lecomte, J. Tailleur, J. Stat. Phys. **145**, 787 (2011)

Appendix A
For Chapter 2

A.1 Derivation of the Path Probability Density (2.4)

In this appendix, we define continuous time Markov processes by taking a continuous time limit in discrete time Markov processes. Then, we derive the path probability density (2.4) in this formulation. Finally, with the obtained path probability density, we show that the probability distribution $P(\boldsymbol{n}, t)$ satisfies the Master equation (2.2).

We consider a finite set Ω, on which a discrete time Markov process is defined. We set an initial time $t = 0$, and fix the initial condition to be $\boldsymbol{n}_0 \in \Omega$. Then after each time interval Δt, we determine if the state $\boldsymbol{n}$ jumps to another state $\boldsymbol{n}'$ or remains in the same state $\boldsymbol{n}$, according to the transition probability

$$\mathrm{Prob}_{\Delta t}(\boldsymbol{n}'|\boldsymbol{n}) = \delta_{\boldsymbol{n}',\boldsymbol{n}}\,[1 - \lambda(\boldsymbol{n})\Delta t] + w(\boldsymbol{n} \to \boldsymbol{n}')\Delta t, \tag{A.1}$$

where $w(\boldsymbol{n} \to \boldsymbol{n}')$ is the transition rate introduced in Sect. 2.2.1, and $\lambda(\boldsymbol{n})$ is the escape rate defined in (2.1). This means that the state remains in the same state with a probability $\lambda(\boldsymbol{n})\Delta t$, and on the other hand, when the system jumps, the place after the jump is determined by a probability that is proportional to $w(\boldsymbol{n} \to \boldsymbol{n}')$. After a certain time t, we have the trajectory of the state, ω, which can be specified by the total number of the jumps n, a collection of transition times $(t_i)_{i=1}^n$ ($t_i = k\Delta t$ with some integers k), and a sequence of states $(\boldsymbol{n}_i)_{i=0}^n$, where $\boldsymbol{n}_i = \boldsymbol{n}(t)$ for $t_i \le t \le t_{i+1}$ with $t_0 = 0$, $t_{n+1} = t$. See Fig. 2.1 for the schematic diagram. From the definition of the transition probability (A.1), the path probability $\mathrm{Prob}(\omega|\boldsymbol{n}_0)$ is calculated as

$$\mathrm{Prob}(\omega|\boldsymbol{n}_0) = [1 - \lambda(\boldsymbol{n}_0)\Delta t]^{\frac{t_1}{\Delta t}} \prod_{i=1}^{N} \left[w(\boldsymbol{n}_{i-1} \to \boldsymbol{n}_i)\Delta t\,[1 - \lambda(\boldsymbol{n}_i)\Delta t]^{\frac{t_{i+1}-t_i}{\Delta t}} \right] \tag{A.2}$$

Now we take $\Delta t \to 0$ limit in this formulation. The obtained model is a continuous time Markov process. The path probability density $P(\omega|\boldsymbol{n}_0)$ is obtained from (A.2) as

T. Nemoto, *Phenomenological Structure for the Large Deviation Principle in Time-Series Statistics*, Springer Theses,
DOI 10.1007/978-981-287-811-3

$$P(\omega|\boldsymbol{n}_0) \equiv \lim_{\Delta t \to 0} \frac{\text{Prob}(\omega|\boldsymbol{n}_0)}{\Delta t^N} = e^{-\lambda(\boldsymbol{n}_0)t_1} \prod_{i=1}^{N} \left[w(\boldsymbol{n}_{i-1} \to \boldsymbol{n}_i) e^{-\lambda(\boldsymbol{n}_i)(t_{i+1}-t_i)} \right], \tag{A.3}$$

which corresponds to (2.3).

Lastly, we derive the master equation (2.2) from this formulation. By using the path probability density (2.3), the distribution function of $\boldsymbol{n}(t)$ is given as

$$\begin{aligned} \left\langle \delta_{\boldsymbol{n}(t),\boldsymbol{n}} \right\rangle = & \sum_{N=0}^{\infty} \sum_{\boldsymbol{n}_0,\boldsymbol{n}_1,\ldots,\boldsymbol{n}_N} \int_0^t dt_N \int_0^{t_N} dt_{N-1} \cdots \int_0^{t_2} dt_1 P_0(\boldsymbol{n}_0) \\ & \delta_{\boldsymbol{n}(t),\boldsymbol{n}} e^{-\int_0^t d\tilde{t} \lambda(\boldsymbol{n}(\tilde{t}))} \prod_{i=1}^{N} \left[w(\boldsymbol{n}_{i-1} \to \boldsymbol{n}_i) \right]. \end{aligned} \tag{A.4}$$

Then, by using $\delta_{\boldsymbol{n}(t),\boldsymbol{n}}$, we rewrite this as follows:

$$\begin{aligned} \left\langle \delta_{\boldsymbol{n}(t),\boldsymbol{n}} \right\rangle = & \sum_{N=0}^{\infty} \sum_{\boldsymbol{n}_0,\boldsymbol{n}_1,\ldots,\boldsymbol{n}_{N-1}} \int_0^t dt_N \int_0^{t_N} dt_{N-1} \cdots \int_0^{t_2} dt_1 e^{-\int_0^t d\tilde{t} \lambda(\boldsymbol{n}(\tilde{t}))} P_0(\boldsymbol{n}_0) \\ & \times \prod_{i=1}^{N-1} \left[w(\boldsymbol{n}_{i-1} \to \boldsymbol{n}_i) \right] w(\boldsymbol{n}_{N-1} \to \boldsymbol{n}) \end{aligned} \tag{A.5}$$

Finally, we take the derivative of this expression with respect to t. By noticing that the derivative is taken at the end-time of the integral with respect to t_N and at the exponential of the time-integral of the escape rate, and by calculating a little bit after that, we obtain

$$\frac{\partial \left\langle \delta_{\boldsymbol{n}(t),\boldsymbol{n}} \right\rangle_{\boldsymbol{n}_0}}{\partial t} = \sum_{\boldsymbol{n}'} \left\langle \delta_{\boldsymbol{n}(t),\boldsymbol{n}'} \right\rangle_{\boldsymbol{n}_0} w(\boldsymbol{n}' \to \boldsymbol{n}) - \lambda(\boldsymbol{n}) \left\langle \delta_{\boldsymbol{n}(t),\boldsymbol{n}} \right\rangle_{\boldsymbol{n}_0}, \tag{A.6}$$

which is the Master equation (2.2).

A.2 Derivation of the Variational Principle (2.29) and (2.30) from Donsker–Varadhan Formula

Here, we derive the variational principle (2.29) and (2.30) from Donsker–Varadhan formula. This derivation was given by C. Maes in a private discussion in Nordita, Stockholm.

A.2.1 Donsker–Varadhan Functional

We first introduce Donsker–Varadhan formula [2]. For a given path ω, we define an empirical measure $\hat{\rho}(\boldsymbol{n};\omega)$ by

$$\hat{\rho}(\boldsymbol{n},\omega) \equiv \frac{1}{t}\int_0^t ds \delta_{\boldsymbol{n}(s),\boldsymbol{n}}. \tag{A.7}$$

In the limit $t\to\infty$, due to the law of large numbers, $\hat{\rho}(\boldsymbol{n},t)$ is equal to the stationary distribution function with probability 1. For the large but finite time t, $\hat{\rho}(\boldsymbol{n},t)$ takes almost the same form as the stationary distribution function, it certainly deviates from it. The probability distribution of this deviation is given as a large deviation principle. Donsker and Varadhan proved it for general Markov dynamics with a mathematically rigorous manner, and derived a formula determining the large deviation functional: For large t, the probability of $\hat{\rho}(\boldsymbol{n},t)$ satisfies a large deviation principle

$$\mathrm{Prob}[\hat{\rho}(\boldsymbol{n},t)=\rho(\boldsymbol{n})] \sim e^{-tI[\rho(\boldsymbol{n})]}. \tag{A.8}$$

with a large deviation functional

$$I[\rho] = -\min_{\tilde{\phi}>0}\sum_{\boldsymbol{n},\boldsymbol{n}'}\rho(\boldsymbol{n})\frac{\left[(w(\boldsymbol{n}\to\boldsymbol{n}')-\lambda(\boldsymbol{n})\delta_{\boldsymbol{n},\boldsymbol{n}'}\right]\tilde{\phi}(\boldsymbol{n}')}{\tilde{\phi}(\boldsymbol{n})}, \tag{A.9}$$

where $\tilde{\phi}(\boldsymbol{n})$ is a variational parameter (vector), where each component takes a positive value. By defining a potential $\tilde{V}(\boldsymbol{n})$ as $\tilde{V}(\boldsymbol{n}) = -2\log\tilde{\phi}(\boldsymbol{n})$, we rewrite it in terms of the modified transition rate $\tilde{w}_h^{\tilde{V}}$ and the corresponding escape rate $\tilde{\lambda}_h^{\tilde{V}}$ introduced in (2.28). The result is

$$I[\rho] = -\min_{\tilde{V}}\sum_{\boldsymbol{n}}\rho(\boldsymbol{n})\left[\tilde{\lambda}_0^{\tilde{V}}(\boldsymbol{n})-\lambda(\boldsymbol{n})\right], \tag{A.10}$$

Below, we show that the variational formula (2.29) and (2.30) are derived from this Donsker–Varadhan formula.

A.2.2 Correspondence Between Biased Ensemble and Modified System

The key idea to derive (2.29) and (2.30) is to construct a correspondence between the biased ensemble and the modified system. We denote the path probability density of $\tilde{w}_h^{\tilde{V}}$-system by $P_h^{\tilde{V}}(\omega|\boldsymbol{n}_0)$ with a given initial condition $\boldsymbol{n}_0$. The path probability density is given as

$$\begin{aligned}
P_h^{\tilde{V}}(\omega|\boldsymbol{n}_0) &= e^{-\int_0^t d\tilde{t}\tilde{\lambda}_h^{\tilde{V}}(\boldsymbol{n}(\tilde{t}))} \prod_{i=1}^N [w(\boldsymbol{n}_{i-1} \to \boldsymbol{n}_i)] \prod_{i=1}^N \left[e^{h\alpha(\boldsymbol{n}_{i-1}\to\boldsymbol{n}_i)-(1/2)\tilde{V}(\boldsymbol{n}_i)+(1/2)\tilde{V}(\boldsymbol{n}_{i-1})} \right] \\
&= e^{htA(\omega)-(1/2)\tilde{V}(\boldsymbol{n}_N)+(1/2)\tilde{V}(\boldsymbol{n}_0)-\int_0^t d\tilde{t}\tilde{\lambda}_h^{\tilde{V}}(\boldsymbol{n}(\tilde{t}))} \prod_{i=1}^N [w(\boldsymbol{n}_{i-1} \to \boldsymbol{n}_i)] \\
&= e^{htA(\omega)-(1/2)\tilde{V}(\boldsymbol{n}_N)+(1/2)\tilde{V}(\boldsymbol{n}_0)-\int_0^t d\tilde{t}\left[\tilde{\lambda}_h^{\tilde{V}}(\boldsymbol{n}(\tilde{t}))-\lambda(\boldsymbol{n}(\tilde{t}))\right]} P(\omega|\boldsymbol{n}_0).
\end{aligned} \tag{A.11}$$

We then neglect $-(1/2)\tilde{V}(\boldsymbol{n}_N) + (1/2)\tilde{V}(\boldsymbol{n}_0)$, since these terms are not proportional to t. Finally, by expressing the time integral with the empirical measure $\hat{\rho}(\boldsymbol{n}, \omega)$, we arrive at

$$e^{htA(\omega)} P(\omega|\boldsymbol{n}_0) \simeq P_h^{\tilde{V}}(\omega|\boldsymbol{n}_0) e^{t \sum_{\boldsymbol{n}} \hat{\rho}(\boldsymbol{n},\omega)\left[\tilde{\lambda}_h^{\tilde{V}}(\boldsymbol{n})-\lambda(\boldsymbol{n})\right]} \tag{A.12}$$

for large t.

A.2.3 Derivation of (2.29) and (2.30)

We set $\tilde{V}$ to be 0 in (A.12). Then, we substitute it into the definition of the cumulant generating function (2.8). We consider the Donsker–Varadhan formula in $\tilde{w}_h^0$-system, for which we denote the Donsker–Varadhan functional in this system by $I_h^0[\rho]$. By using $I_h^0[\rho]$ in the obtained expression and evaluating it by using a saddle-point method, we obtain the cumulant generating function as

$$G(h) = \max_{\tilde{\rho}} \left[\sum_{\boldsymbol{n}} \tilde{\rho}(\boldsymbol{n}) \left(\tilde{\lambda}_h^0(\boldsymbol{n}) - \lambda(\boldsymbol{n}) \right) - I_h^0[\tilde{\rho}] \right]. \tag{A.13}$$

Here, we recall the expression of the Donsker–Varadhan formula (A.9). We denote by V_h^* the optimal $\tilde{V}$ that appeared in the variational principle in the Donsker–Varadhan formula:

$$V_h^* = \operatorname*{argmin}_{\tilde{V}} \left[\sum_{\boldsymbol{n}} \rho(\boldsymbol{n}) \left(\tilde{\lambda}_h^{\tilde{V}}(\boldsymbol{n}) - \lambda_h^0(\boldsymbol{n}) \right) \right]. \tag{A.14}$$

By combining this definition with (A.13), we thus obtain

$$G(h) = \max_{\tilde{\rho}} \left[\sum_{\boldsymbol{n}} \tilde{\rho}(\boldsymbol{n}) \left[\tilde{\lambda}_h^{V_h^*}(\boldsymbol{n}) - \lambda(\boldsymbol{n}) \right] \right]. \tag{A.15}$$

The obtained formula (A.15) is basically the same as (2.30). The only difference in it is the variational parameter $\tilde{\rho}$. As the final step, we replace this variational parameter $\tilde{\rho}$ by a potential. First, we consider the variational equation for the minimization problem (A.14), which determines $V_h^*(\boldsymbol{n})$. The equation is

$$\sum_{\boldsymbol{n},\boldsymbol{n}'} \tilde{\rho}(\boldsymbol{n}) \left[\delta\tilde{V}(\boldsymbol{n}) - \delta\tilde{V}(\boldsymbol{n}')\right] \tilde{w}_h^{V_h^*}(\boldsymbol{n} \to \boldsymbol{n}') = 0. \tag{A.16}$$

It is further rewritten as

$$\sum_{\boldsymbol{n}} \delta\tilde{V}(\boldsymbol{n}) \sum_{\boldsymbol{n}'} \left[\tilde{\rho}(\boldsymbol{n})\tilde{w}_h^{\tilde{V}_h^*}(\boldsymbol{n} \to \boldsymbol{n}') - \tilde{\rho}(\boldsymbol{n}')\tilde{w}_h^{V_h^*}(\boldsymbol{n}' \to \boldsymbol{n})\right] = 0. \tag{A.17}$$

Since $\delta\tilde{V}$ is arbitrary, we obtain

$$\sum_{\boldsymbol{n}'} \left[\tilde{\rho}(\boldsymbol{n})\tilde{w}_h^{V_h^*}(\boldsymbol{n} \to \boldsymbol{n}') - \tilde{\rho}(\boldsymbol{n}')\tilde{w}_h^{V_h^*}(\boldsymbol{n}' \to \boldsymbol{n})\right] = 0. \tag{A.18}$$

It means that $V_h^*(\boldsymbol{n})$ is determined in such a way that $\tilde{\rho}(\boldsymbol{n})$ is the stationary probability for the modified system with $\tilde{w}_h^{V_h^*}$. (For to the uniqueness and existence, see Proposition III.1 in Ref. [1].) Since the stationary probability ρ_h^V is determined uniquely for a given $\tilde{w}_h^V$, we find the one-to-one correspondence between V_h^* and $\tilde{\rho}$. Thus, we may rewrite (A.15) as

$$G(\sigma) = \max_V \left[\sum_x \mu_\sigma^V(x) \left(D_\sigma^V(x) - D(x)\right)\right]. \tag{A.19}$$

This is exactly the same as (2.30), which directly leads to (2.29).

A.3 Derivation of (2.36)

Here, we show that $\left\langle \delta_{\boldsymbol{n}(t),\boldsymbol{n}} e^{htA(\omega)} \right\rangle_{\boldsymbol{n}_0}$ satisfies the same equations as (2.34) and (2.35), which leads to (2.36). For the derivation of (2.34), by substituting t in $\left\langle \delta_{\boldsymbol{n}(t),\boldsymbol{n}} e^{htA(\omega)} \right\rangle_{\boldsymbol{n}_0}$ by 0, we can easily check $\left\langle \delta_{\boldsymbol{n}(t),\boldsymbol{n}} e^{htA(\omega)} \right\rangle_{\boldsymbol{n}_0} = \delta_{\boldsymbol{n},\boldsymbol{n}_0}$. On the other hand, for (2.35), we start with the path probability density (2.4). From the expression, $\left\langle \delta_{\boldsymbol{n}(t),\boldsymbol{n}} e^{htA(\omega)} \right\rangle_{\boldsymbol{n}_0}$ is written as

$$\begin{aligned} \left\langle \delta_{\boldsymbol{n}(t),\boldsymbol{n}} e^{htA(\omega)} \right\rangle_{\boldsymbol{n}_0} = & \sum_{N=0}^{\infty} \sum_{\boldsymbol{n}_1,\boldsymbol{n}_2,\ldots,\boldsymbol{n}_N} \int_0^t dt_N \int_0^{t_N} dt_{N-1} \cdots \int_0^{t_2} dt_1 \\ & \delta_{\boldsymbol{n}(t),\boldsymbol{n}} e^{htA(\omega)} e^{-\int_0^t d\tilde{t}\lambda(\boldsymbol{n}(\tilde{t}))} \prod_{i=1}^{N} \left[w(\boldsymbol{n}_{i-1} \to \boldsymbol{n}_i)\right]. \end{aligned} \tag{A.20}$$

Then, in this expression, we make $\delta_{\boldsymbol{n}(t),\boldsymbol{n}} e^{htA(\omega)}$ included inside the path-probability. The result is

$$\left\langle \delta_{\boldsymbol{n}(t),\boldsymbol{n}} e^{htA(\omega)} \right\rangle_{\boldsymbol{n}_0} = \sum_{N=0}^{\infty} \sum_{\boldsymbol{n}_1,\boldsymbol{n}_2,\ldots,\boldsymbol{n}_{N-1}} \int_0^t dt_N \int_0^{t_N} dt_{N-1} \cdots \int_0^{t_2} dt_1 e^{-\int_0^t d\tilde{t} \lambda(\boldsymbol{n}(\tilde{t}))}$$
$$\times \prod_{i=1}^{N-1} \left[w(\boldsymbol{n}_{i-1} \to \boldsymbol{n}_i) e^{h\alpha(\boldsymbol{n}_{i-1} \to \boldsymbol{n}_i)} \right] w(\boldsymbol{n}_{N-1} \to \boldsymbol{n}) e^{h\alpha(\boldsymbol{n}_{N-1} \to \boldsymbol{n})} \tag{A.21}$$

We differentiate this expression with respect to t. The differentiation is taken at the end-time of the integral with respect to t_N and at the exponential of the time-integral of the escape rate. After some calculations, we arrive at

$$\frac{\partial \left\langle \delta_{\boldsymbol{n}(t),\boldsymbol{n}} e^{htA(\omega)} \right\rangle_{\boldsymbol{n}_0}}{\partial t}$$
$$= \sum_{\boldsymbol{n}'} \left\langle \delta_{\boldsymbol{n}(t),\boldsymbol{n}'} e^{htA(\omega)} \right\rangle_{\boldsymbol{n}_0} w(\boldsymbol{n}' \to \boldsymbol{n}) e^{h\alpha(\boldsymbol{n}' \to \boldsymbol{n})} - \lambda(\boldsymbol{n}) \left\langle \delta_{\boldsymbol{n}(t),\boldsymbol{n}} e^{htA(\omega)} \right\rangle_{\boldsymbol{n}_0}, \tag{A.22}$$

which is the corresponding equation to (2.35).

A.4 Derivation of the Theoretical Basis (2.48) of the Rare-Event Sampling Method

We denote by $\phi(\boldsymbol{n}; h)$ the left-eigenvector corresponding to the largest eigenvalue of the matrix $L^h_{\boldsymbol{n}',\boldsymbol{n}}$ defined in (2.25). In this appendix, then, we derive

$$\phi(\boldsymbol{n}; l\delta h) \propto \prod_{k=0}^{l-1} \left\langle e^{\tau \delta h A(\omega)} \right\rangle_{\boldsymbol{n}}^{k\delta h} \tag{A.23}$$

for $l = 1, 2, 3, \ldots$ From this relation, it is easy to check (2.48) follows. Indeed, by combining this relation with the argument in Sect. 2.2.5, we obtain (2.48).

First, we define

$$w^h(\boldsymbol{n} \to \boldsymbol{n}') = w(\boldsymbol{n} \to \boldsymbol{n}') e^{h\alpha(\boldsymbol{n} \to \boldsymbol{n}')} \frac{\phi(\boldsymbol{n}'; l\delta h)}{\phi(\boldsymbol{n}; l\delta h)}. \tag{A.24}$$

Then, by using $w^h(\boldsymbol{n} \to \boldsymbol{n}')$, we define a matrix

$$L^{h,h'}_{\boldsymbol{n},\boldsymbol{n}'} \equiv w^{h'}(\boldsymbol{n}' \to \boldsymbol{n}) e^{h\alpha(\boldsymbol{n}' \to \boldsymbol{n})} - \lambda^{w^{h'}}(\boldsymbol{n}) \delta_{\boldsymbol{n},\boldsymbol{n}'}, \tag{A.25}$$

where $\lambda^{w^{h'}}(\boldsymbol{n}) \equiv \sum_{\boldsymbol{n}'} w^{h'}(\boldsymbol{n} \to \boldsymbol{n}')$. Let $K^{h,h'}$ and $\phi^{h,h'}$ be the largest eigenvalue and the corresponding left-eigenvector of (A.25). Then, we can prove the following *multiplicative property* for the eigenvector and *additive property* for the eigenvalue of the matrix (A.25):

$$\phi^{h+h',0} = \phi^{h,h'}\phi^{h',0}, \tag{A.26}$$

$$K^{h+h',0} = K^{h,h'} + K^{h',0}. \tag{A.27}$$

- Proof:
  First, we write the eigenvalue equations for $\phi^{h+h',0}$, $\phi^{h',0}$, and $\phi^{h,h'}$. Those are

$$\sum_{\boldsymbol{n}'} w(\boldsymbol{n} \to \boldsymbol{n}')e^{(h+h')\alpha(\boldsymbol{n}\to\boldsymbol{n}')}\frac{\phi^{h+h',0}(\boldsymbol{n}')}{\phi^{h+h',0}(\boldsymbol{n})} - \lambda(\boldsymbol{n}) = K^{h+h',0}, \tag{A.28}$$

$$\sum_{\boldsymbol{n}'} w(\boldsymbol{n} \to \boldsymbol{n}')e^{h'\alpha(\boldsymbol{n}\to\boldsymbol{n}')}\frac{\phi^{h',0}(\boldsymbol{n}')}{\phi^{h',0}(\boldsymbol{n})} - \lambda(\boldsymbol{n}) = K^{h',0}, \tag{A.29}$$

$$\begin{aligned}\sum_{\boldsymbol{n}'} w(\boldsymbol{n} \to \boldsymbol{n}')e^{(h+h')\alpha(\boldsymbol{n}\to\boldsymbol{n}')}\frac{\phi^{h,h'}(\boldsymbol{n}')\phi^{h',0}(\boldsymbol{n}')}{\phi^{h,h'}(\boldsymbol{n})\phi^{h',0}(\boldsymbol{n})} - \lambda^{u^{h'}}(\boldsymbol{n})\\ = K^{h,h'}.\end{aligned} \tag{A.30}$$

We sum up (A.29) and (A.30). Since the first term of (A.29) is the same as the second one in (A.30), these terms cancel each other. The result is

$$\begin{aligned}\sum_{\boldsymbol{n}'} w(\boldsymbol{n} \to \boldsymbol{n}')e^{(h+h')\alpha(\boldsymbol{n}\to\boldsymbol{n}')}\frac{\phi^{h,h'}(\boldsymbol{n}')\phi^{h',0}(\boldsymbol{n}')}{\phi^{h,h'}(\boldsymbol{n})\phi^{h',0}(\boldsymbol{n})} - \lambda(\boldsymbol{n})\\ = K^{h,h'} + K^{h',0}.\end{aligned} \tag{A.31}$$

From the Perron-Frobenius theory for irreducible matrices [3], the eigenvector of $L^{h,0}_{\boldsymbol{n},\boldsymbol{n}'}$ that takes positive value is unique, and the corresponding eigenvalue should become the largest one $G^{h+h',0}$. Thus, by comparing (A.28) with (A.31), we obtain (A.26) and (A.27).

Next, we denote by $\langle\ \rangle^{w^h}$ the expected value in the stationary state of the system generated by the transition rate $w^h(\boldsymbol{n} \to \boldsymbol{n}')$. We will show that this expected value and the expected value $\langle\ \rangle^h$ defined in Sect. 2.3.4 are equal. Meanwhile, we show a formula for $\langle\ \rangle^{w^h}$ and finally we will show the equivalence. First, we have a relation

$$\phi^{h,h'}(\boldsymbol{n}) \propto \left\langle e^{thA(\omega)}\right\rangle^{w^{h'}}_{\boldsymbol{n}} \tag{A.32}$$

for large $t(> t_a)$. The derivation is the same as the one for (2.38), so we don't repeat it here. Then, by combining (A.32) with (A.28), we obtain

$$\phi^{h+h',0}(\boldsymbol{n}) \propto \left\langle e^{thA(\omega)}\right\rangle^{w^{h'}}_{\boldsymbol{n}} \phi^{h',0}(\boldsymbol{n}), \tag{A.33}$$

or equivalently,

$$\phi^{h+h',0}(\boldsymbol{n}) \propto \left\langle e^{thA(\omega)} \right\rangle_{\boldsymbol{n}}^{w^{h'}} \left\langle e^{th'A(\omega)} \right\rangle_{\boldsymbol{n}}^{0}, \quad \text{(A.34)}$$

where we used (A.32). The equivalence between $\langle\ \rangle^{w^h}$ and $\langle\ \rangle^h$ is easily checked. Indeed, by using (2.38) in the definition of $\langle\ \rangle^h$, the relation follows. We thus have

$$\phi^{h+h',0}(\boldsymbol{n}) \propto \left\langle e^{thA(\omega)} \right\rangle_{\boldsymbol{n}}^{h'} \left\langle e^{th'A(\omega)} \right\rangle_{\boldsymbol{n}}^{0}. \quad \text{(A.35)}$$

The generalisation to the case, where a given h is divided into l pieces, is straightforward. Therefore, we arrive at (A.23).

A.5 Application of Our Method to Obtain a L Dependence of $\tilde{G}(\tilde{h})$ in (2.72)

For investigating the singular behaviour of $G(h)$ in greater detail, a scaled biasing parameter $\tilde{h} \equiv hL$ has been used [4, 5]. The problem was in the numerical study of it because the population dynamics method does not exhibit good convergence of $\tilde{G}(\tilde{h})$ for relatively large L [5]. In this appendix, we show that our method can also be applied to obtain the reliable L dependence of $\tilde{G}(\tilde{h})$ even in this case.

A.5.1 (i) Sufficiently Large Truncating Number *r* for Severe L

We investigate r dependence of $\tilde{G}(\tilde{h})$ for several L. We will conclude that $r = 4$ is sufficiently large to obtain $\tilde{G}(\tilde{h}) \equiv G(L^{-1}\tilde{h})$ for $\tilde{h} < 0$. First, we define $\tilde{G}(\tilde{h}, r)$ as the obtained cumulant generating function with the truncating number r. Then, we show the numerical examples of $\tilde{G}(\tilde{h}, r)$ in Fig. A.1. In the figure, we plot $\tilde{G}(\tilde{h}, 1)$, $\tilde{G}(\tilde{h}, 2)$, $\tilde{G}(\tilde{h}, 3)$, and $\tilde{G}(\tilde{h}, 4)$ for $L = 10$ (i), 20 (ii), 30 (iii), and 60 (iv). In the same figure, we also plot the straight line (Ex) with the slope $4c^2(1 - c)$, which is the expected value of the activity in the unmodified system ($\tilde{h} = 0$). We note that the straight line corresponds to $\tilde{G}(\tilde{h}, 0)$, because $r = 0$ means that there are no modifications. From the figure, we find that the differences between $\tilde{G}(\tilde{h}, 3)$ and $\tilde{G}(\tilde{h}, 4)$ are small even for larger L (say $L = 20$, 30, and 60).

We then quantitatively evaluate those small differences. For this, we introduce a difference function

$$\delta\tilde{G}(\tilde{h}, r) = \tilde{G}(\tilde{h}, r + 1) - \tilde{G}(\tilde{h}, r). \quad \text{(A.36)}$$

Then, in Fig. A.2, we plot the logarithm of $\delta G(r, h)$ as a function of r for $\tilde{h} = -0.225, -0.45, -0.675$ and -0.9 with $L = 20$ (i), 30 (ii), and 60 (iii). We can see

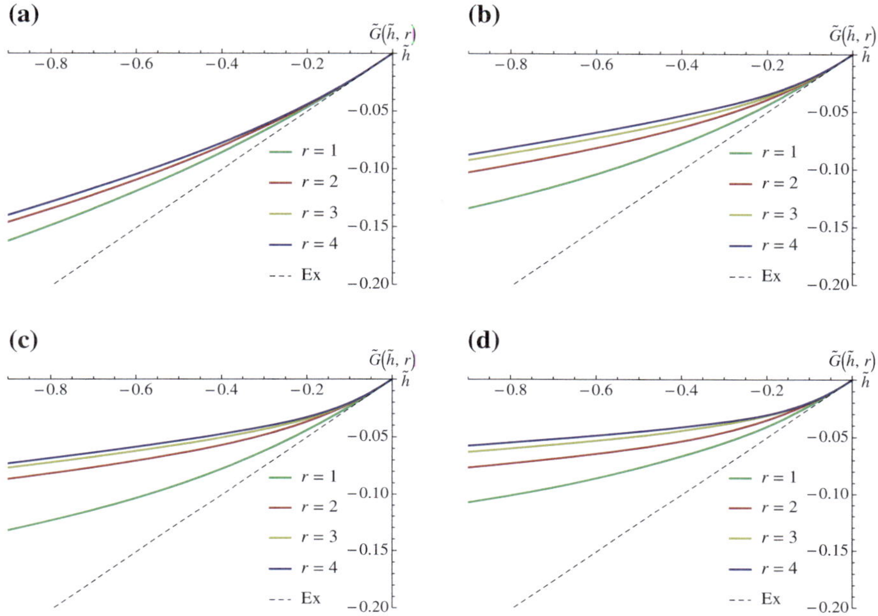

Fig. A.1 $\tilde{G}(\tilde{h}, r)$ for $r = 1, 2, 3, 4$ with $L = 10$ (**a**), 20 (**b**), 30 (**c**), and 60 (**d**), where we set $c = 0.3$. We also plot the *straight line* of slope $4c^2(1 - c)$ (Ex) in each figure, which corresponds to $\tilde{G}(\tilde{h}, 0)$

the linear dependence of $\log[\delta G(r, h)]$ on r. This means exponentially fast decay of $\delta\tilde{G}$, which may indicate that larger r isn't needed to obtain the correct $\tilde{G}(\tilde{h})$.

Finally, we evaluate the error due to the truncation of $r = 4$. From Fig. A.2, we assume that the decaying of $\delta\tilde{G}(\tilde{h}, r)$ with r is well described by an exponential function:

$$\delta\tilde{G}_{\text{lin}}(\tilde{h}, r) = e^{a(\tilde{h})r + b(\tilde{h})}, \tag{A.37}$$

where $a(\tilde{h})$ and $b(\tilde{h})$ are coefficients determined from the least squares fit of data points $\delta\tilde{G}(\tilde{h}, r)$. The examples of the least squares fit are shown in Fig. A.2. By using this difference function, we define an (exponential decaying) approximation function of $\tilde{G}(\tilde{h}, r)$ as

$$\tilde{G}_{\text{lin}}(\tilde{h}, r) \equiv \tilde{G}(\tilde{h}, 4) + \sum_{s=4}^{r-1} \delta\tilde{G}_{\text{lin}}(\tilde{h}, s) \tag{A.38}$$

for $r = 5, 6, \ldots$ Especially, here, we denote $\tilde{G}_{\text{lin}}(\tilde{h}, L/2 - 1)$ by $\tilde{G}_{\text{lin}}(\tilde{h})$:

$$\tilde{G}_{\text{lin}}(\tilde{h}) \equiv \tilde{G}_{\text{lin}}(\tilde{h}, L/2 - 1). \tag{A.39}$$

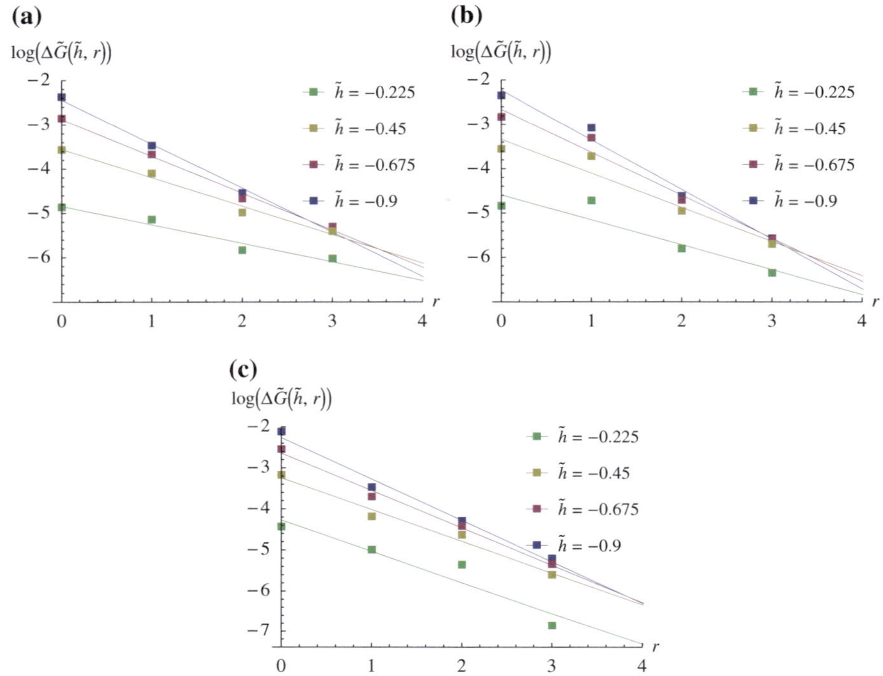

Fig. A.2 The logarithm of the difference function $\delta\tilde{G}(\tilde{h}, r)$ for $r = 0, 1, 2,$ and 3 with $L = 20$ (**a**), 30 (**b**), and 60 (**c**). We set $c = 0.3$. We also plot *straight lines* obtained from a least squares fit of those data points

Since $\tilde{G}(\tilde{h}, r)$ with $r \simeq L/2$ is equal to $\tilde{G}(\tilde{h})$, we regard $\tilde{G}_{\mathrm{lin}}(\tilde{h})$ as an approximation function of $\tilde{G}(\tilde{h})$. We plot $\tilde{G}_{\mathrm{lin}}(\tilde{h})$ and $\tilde{G}(\tilde{h}, r)$ for $r = 1, 2, 3, 4$ in Fig. A.3. The figure shows that the differences between $\tilde{G}_{\mathrm{lin}}(\tilde{h})$ and $\tilde{G}(\tilde{h}, 4)$ are quite small even for large L. Therefore, we judge that $r = 4$ is sufficiently large to obtain $\tilde{G}(\tilde{h})$ even for those large L.

A.5.2 (ii) L Dependence of $\tilde{G}(\tilde{h})$ Obtained from Truncation

In the previous subsection, we judged that $r = 4$ was sufficiently large to obtain $\tilde{G}(\tilde{h})$. Here, by using this result, we show the L dependence of $\tilde{G}(\tilde{h})$. In Fig. A.4, we plot $\tilde{G}(\tilde{h})$ with $r = 4$ for various values of L . Even though it was conjectured that $\tilde{G}(\tilde{h})$ has a non-differentiable point in the limit $L \to \infty$ [4], our result does not show any clear sign of such a cusp. More and more larger system sizes are required for investigating the singular behaviour of $\tilde{G}(\tilde{h})$.

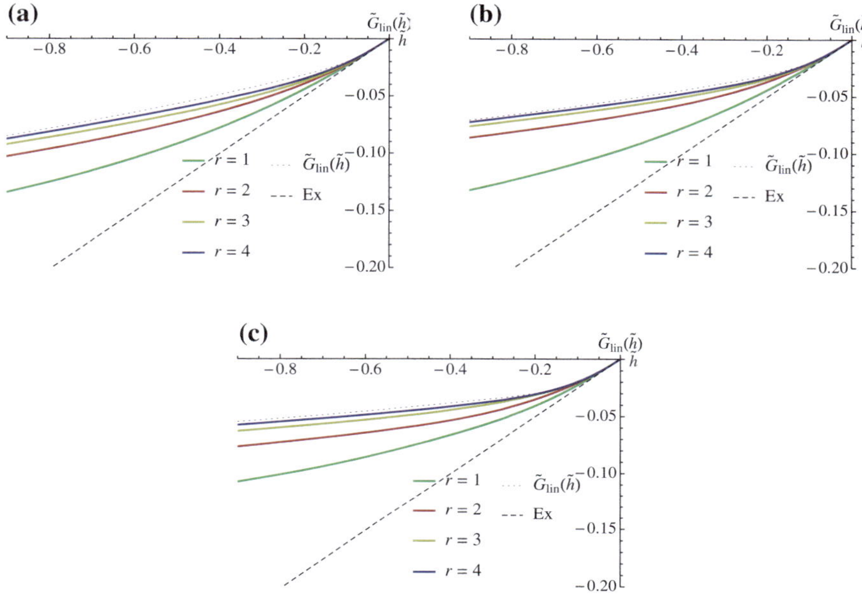

Fig. A.3 $\tilde{G}(\tilde{h}, r)$ and $\tilde{G}_{\text{lin}}(\tilde{h})$ for $r = 1, 2, 3, 4$ with $L = 20$ (**a**), 30 (**b**), and 60 (**c**), where we set $c = 0.3$

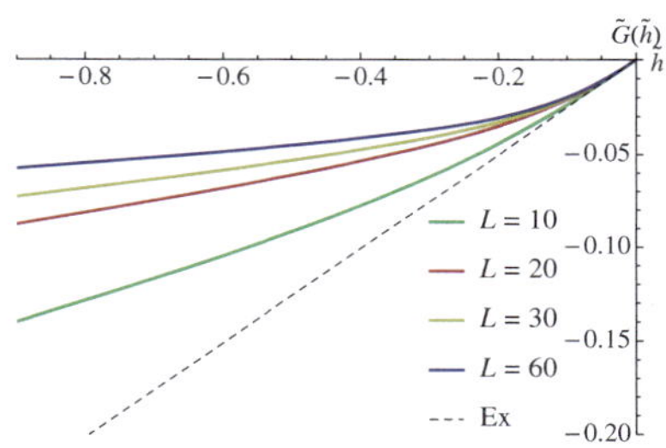

Fig. A.4 $\tilde{G}(\tilde{h})$ for various values of L with $r = 4$ fixed. We also plot the *straight line* (Ex) of slope $4c^2(1 - c)$ that corresponds to the expected value of the activity in the unbiased system ($h = 0$)

References

[1] C. Maes, K. Netočný, B. Wynants, J. Phys. A: Math. Theor. **45**, 455001 (2012)
[2] M.D. Donsker, S.R. Varadhan, Commun. Pure Appl. Math. **28**, 1 (1975)
[3] E. Seneta, *Non-Negative Matrices and Markov Chains*, 2nd edn. (Springer, New York, 2006)
[4] T. Bodineau, C. Toninelli, Commun. Math. Phys. **311**, 357 (2012)
[5] T. Bodineau, V. Lecomte, C. Toninelli, J. Stat. Phys. **147**, 1 (2012)

Appendix B
For Chapter 3

B.1 Derivation of Finite-Size Corrections to Modifying Free Energy $\Delta f_{s_c}^{(1)}(\rho)$ in (3.55) and (3.56)

In this appendix, we derive the finite-size correction $\Delta f_{s_c}^{(1)}(\rho)$ given in (3.55) and (3.56).

B.1.1 For the Region $\rho > \rho_c^L$

We first focus on the region $\rho > \rho_c^L$, which corresponds to (3.56). In this region, from the expression of the leading order of the free energy (3.50), we know that $\phi(n)$ doesn't satisfy any large deviation principle. We thus define $\tilde{\phi}(\rho) = \phi(\rho L)$ and assume the differentiability of it:

$$\tilde{\phi}(\rho \pm 1/L) = \tilde{\phi}(\rho) \pm \frac{\partial \tilde{\phi}}{\partial \rho}\frac{1}{L} + O(1/L^2). \tag{B.1}$$

By using this scaling property, we rewrite the left-hand side of (3.13). The result is

$$\tilde{\phi}(\rho)\left\{-\tilde{s}_c\rho\left[c + (1-2c)\rho\right] + c\right\} + \frac{\partial\tilde{\phi}(\rho)}{\partial \rho}\rho(c-\rho) + O(1/L^2) = 0, \tag{B.2}$$

where we defined $\tilde{s}_c \equiv sL$. By solving this differential equation, we obtain the modifying free energy with finite-size correction, which we denote by $-2\log\tilde{\phi}$, as follows:

T. Nemoto, *Phenomenological Structure for the Large Deviation Principle in Time-Series Statistics*, Springer Theses,
DOI 10.1007/978-981-287-811-3

$$
\begin{aligned}
&-2\log\tilde{\phi}(\rho)\\
&=-2\left[\tilde{s}_{\rm c}\rho(2c-1)-\log\rho+(-\tilde{s}_{\rm c}2c(1-c)+1)\log|c-\rho|\right]+\text{const.}\\
&=-2\left[\frac{\rho(2c-1)}{2c(1-c)}-\log\rho\right]+\text{const.},
\end{aligned}
\tag{B.3}
$$

where we used a relation $\tilde{s}_{\rm c}=1/(2c(1-c))+O(1/L)$ from the second to the third line. Since the leading term to the modifying free energy is constant as shown in (3.50), we arrive at

$$
\Delta f^{(1)}_{s_{\rm c}}(\rho)=-2\left[\frac{\rho(2c-1)}{2c(1-c)}-\log\rho\right]+\text{const.}
\tag{B.4}
$$

for $\rho>\rho_{\rm c}^L$, which corresponds to (3.56).

B.1.2 For the Region $\rho\le\rho_{\rm c}^L$

Next, we focus on the region $\rho\le\rho_{\rm c}^L$, which corresponds to (3.55). In this region, $\phi(n)$ satisfies a large deviation principle. We thus directly substitute the definition (3.54) with the aide of the result of the leading order give by (3.49) into the eigenvalue equation (3.43). In the calculation, we evaluate $\phi(n+1)/\phi(n)$ as

$$
\begin{aligned}
\frac{\phi(n+1)}{\phi(n)}&=e^{\partial f_{\rm e}/\partial\rho+1/(2L)\partial^2 f_{\rm e}/\partial\rho^2-1/(2L)\partial\Delta f^{(1)}_{s_{\rm c}}/\partial\rho}\\
&=\frac{(1-c)\rho}{c(1-\rho)}e^{(1/(2L\rho(1-\rho)))}e^{-1/(2L)\partial\Delta f^{(1)}_{s_{\rm c}}/\partial\rho}.
\end{aligned}
$$

By using this expression in (3.43), we arrive at a differential equation

$$
\begin{aligned}
\frac{\partial\Delta f^{(1)}_{s_{\rm c}}(\rho)}{\partial\rho}&=\frac{2}{\rho}-2\frac{\frac{c}{\rho}+[(1-c)\rho+c(1-\rho)]\left[\frac{1}{2\rho(1-\rho)}-\frac{1}{2c(1-c)}\right]}{-(1-c)\rho+c(1-\rho)}\\
&=-\frac{1}{\rho}+\frac{1}{1-\rho}-\frac{2}{c-\rho}-\frac{1-2c}{c(1-c)},
\end{aligned}
\tag{B.5}
$$

which leads to

$$
\Delta f^{(1)}_{s_{\rm c}}(\rho)=-\log\frac{\rho(1-\rho)}{(c-\rho)^2}-\frac{\rho(1-2c)}{c(1-c)}+\text{const.}
\tag{B.6}
$$

This corresponds to (3.55).

B.2 Derivation of $a^*(s)$ Given in (3.77)

Here, we derive the expression $a^*(s)$ given as (3.77) by solving (3.74).

In order to evaluate the variational function $\Psi(a)$ given in (3.75), we divide the domain of the summation in (3.75) into three parts:

(i) $n < n_c$

$$\Psi_<(a) \equiv (1/L) \sum_{n<n_c} P^s(n)\left[\tilde{r}(n) - r(n)\right], \tag{B.7}$$

(ii) $n > n_c + 1$

$$\Psi_>(a) \equiv (1/L) \sum_{n>n_c} P^s(n)\left[\tilde{r}(n) - r(n)\right], \tag{B.8}$$

(iii) $n \equiv n_c, n_c + 1$

$$\Psi_=(a) \equiv (1/L) \sum_{n=n_c,n_c+1} P^s(n)\left[\tilde{r}(n) - r(n)\right]. \tag{B.9}$$

The dependence of a in (i) and (ii) is linear because $\tilde{r}(n)$ is an independent function of a. For these part, we define two constants $\Omega_<$ and $\Omega_>$ that doesn't depend on a:

$$\Omega_< = \frac{1}{2L}\left\langle \tilde{r}_i e^{-s} - r\right\rangle_i, \tag{B.10}$$

$$\Omega_> = \frac{1}{2L}\left\langle \tilde{r}_a e^{-s} - r\right\rangle_a, \tag{B.11}$$

where $\tilde{r}_{i,a}(n)$ is defined as

$$\tilde{r}_{i,a}(n) = nc\left(1 - \frac{n}{L}\right)\frac{e^{-Lf_{i,a}((n+1)/L)/2}}{e^{-Lf_{i,a}(n/L)/2}} + n(1-c)\left(\frac{n}{L} - \frac{1}{L}\right)\frac{e^{-Lf_{i,a}((n-1)/L)/2}}{e^{-Lf_{i,a}(n/L)/2}}. \tag{B.12}$$

By using these constants, $\Psi_<(a)$ and $\Psi_>(a)$ are written as

$$\Psi_<(a) = (1+a)\Omega_<, \tag{B.13}$$

$$\Psi_>(a) = (1-a)\Omega_>. \tag{B.14}$$

On the other hand, the dependence of a in (iii) is more complicated. We first write down $\Psi_=(a)$ as

$$\begin{aligned}\Psi_=(a) = &\frac{n_c}{L}c(1 - \frac{n_c}{L})\frac{\phi(n_c+1)}{\phi(n_c)}P^s(n_c)e^{-s} \\ &+ \frac{(n_c+1)}{L}(1-c)\frac{n_c}{L}\frac{\phi(n_c)}{\phi(n_c+1)}P^s(n_c+1)e^{-s} + \cdots,\end{aligned} \tag{B.15}$$

where $\cdots$ represents the terms that are proportional to a. Since this linear dependence of a is exponentially smaller than the one in $\Psi_<(a)$ and $\Psi_>(a)$, we omit this part hereafter. Next, with a relation between $P^s(n)$ and $\phi(n)$

$$P^s(n+1)\frac{\phi(n)}{\phi(n+1)} = P^s(n)\frac{\phi(n+1)P_{\text{eq}}(n+1)}{\phi(n)P_{\text{eq}}(n)}, \tag{B.16}$$

we find that the first term and the second term in the right hand side of (B.15) are equal. Also, by recalling a dependence of $P^s(n)$ and $\phi(n)$, we find that this term is proportional to $\sqrt{(}1-a^2)$. Thus, by defining the coefficient of it, $\Omega_=(a)$, as

$$\Omega_= \equiv 2\frac{n_c}{L}c(1-\frac{n_c}{L})P_{\text{i}}(n_c)\frac{\mathrm{e}^{-Lf_{\text{a}}((n_c+1)/L)/2}}{\mathrm{e}^{-Lf_{\text{i}}(n_c/L)/2}}\mathrm{e}^{-s}, \tag{B.17}$$

we arrive at

$$\Psi_=(a) = \sqrt{1-a^2}\Omega_=. \tag{B.18}$$

The non-linear dependence in $\Psi_=(a)$ is important. Even though $\Psi_=(a)$ is exponentially small compared with the other parts $\Psi_>(a)$, $\Psi_<(a)$, due to the non-liner dependence, we need to consider this term. As seen in the main text, this smallness of $\Psi_=(a)$ is the origin of the exponentially small width of the coexistence region.

From (B.13), (B.14), and (B.18), we obtain

$$\Psi(a) = \Omega_< + \Omega_> + a(\Omega_< - \Omega_>) + \sqrt{1-a^2}\Omega_=. \tag{B.19}$$

By maximising $\Psi(a)$ with respect to a, we finally obtain the expression of $a^*(s)$ as

$$a^*(s) = \frac{A}{\sqrt{1+A^2}}, \tag{B.20}$$

with

$$A = (\Omega_< - \Omega_>)/\Omega_=. \tag{B.21}$$

This is (3.77).

B.3 Analytical Expressions of the Scaling Functions for $\partial G(s)/\partial s$ and $\partial^2 G(s)/\partial s^2$

Here, we derive the scaling function for $\partial G(s)/\partial s$ and $\partial^2 G(s)/\partial s^2$.

We first recall the relation between $\partial G(s)/\partial s$ and the expected values of $\rho = n/L$. As already introduced in Sect. 3.2.1, we have

$$
\begin{aligned}
-\frac{\partial G(s)}{\partial s} &= \sum_n \sum_{n'} P^s(n) w(n \to n') = \sum_n P^s(n)\lambda(n) \\
&= \rho(s)(Lc + c - 1) + \rho(s)^2 L(1 - 2c) + (1 - 2c)\chi(s).
\end{aligned}
\tag{B.22}
$$

Then, by substituting these $\rho(s)$ and $\chi(s)$ by the one in (3.70) and (3.71), changing the variables to x, and using (3.77) and (3.86), we rewrite it as

$$
\begin{aligned}
&-\frac{\partial G(s)}{\partial s}\Big|_{s=s_c+\kappa^{-1}x} \\
&= \frac{1}{2}\left[\langle\rho\rangle_i + \langle\rho\rangle_a\right](Lc + c - 1) + \frac{1-2c}{2} L\left[\langle\rho^2\rangle_i + \langle\rho^2\rangle_a\right] \\
&\quad + \frac{2x\left[\langle\rho_a\rangle - \langle\rho\rangle_i\right]^{-1}}{\sqrt{1 + 4x^2\left[\langle\rho_a\rangle - \langle\rho\rangle_i\right]^{-2}}} \left\{\frac{1}{2}\left[\langle\rho\rangle_i - \langle\rho\rangle_a\right](Lc + c - 1)\right. \\
&\qquad\qquad\qquad\qquad\qquad\qquad \left. + \frac{1-2c}{2} L\left[\langle\rho^2\rangle_i - \langle\rho^2\rangle_a\right].\right\}
\end{aligned}
$$

Then, by taking $L \to \infty$, we arrive at

$$
-\lim_{L\to\infty} \frac{1}{L}\frac{\partial G(s)}{\partial s}\Big|_{s=s_c+\kappa^{-1}x} = c^2(1-c)\left[1 - \frac{2xc^{-1}}{\sqrt{1+4x^2c^{-2}}}\right]. \tag{B.23}
$$

Since the expression of $\partial G(s)/\partial s$ is determined, we can obtain $\partial^2 G(s)/\partial s^2$ just by taking the derivative of it. By noticing that $\partial^2 G(s)/\partial s^2$ is not directly connected to the equilibrium distribution function $P^s(n)$ from the definition, this property is suggestive. We thus obtain

$$
\lim_{L\to\infty} \frac{1}{L\kappa}\frac{\partial^2 \psi(s)}{\partial s^2}\Big|_{s=s_c+\kappa^{-1}x} = 2c(1-c)\frac{1}{(1+4x^2c^{-2})^{3/2}}. \tag{B.24}
$$

B.4 Variational Principle to Determine the Ground State Energy in Quantum Systems

Here, we show that the variational principle (3.21) that gives the cumulant generating function, when the system satisfies detailed balance condition, is reduced to the one for determining the ground energy in quantum systems. The key is a symmetrisation of the matrix $L^s_{n',n}$ in the largest eigenvalue problem (3.20). Thanks to the detailed balance condition, such a symmetrisation is possible. For symmetric matrices, a variational principle for determining the largest (or lowest) eigenvalue problem is well known in quantum mechanics. We apply this variational principle to the system, and show that this formula and (3.21) is equivalent.

The detailed balance condition is given as (3.7). By using this condition, we rewrite $L^s_{n',n}$ as

$$\begin{aligned} L^s_{n',n} &= w(n \to n')e^{-s} - \delta_{n,n'}\lambda(n) \\ &= P_{\rm eq}(n')w(n' \to n)P_{\rm eq}(n)^{-1}e^{-s} - \delta_{n,n'}\lambda(n) \\ &= P_{\rm eq}(n')^{1/2}\left[P_{\rm eq}(n')^{1/2}w(n' \to n)P_{\rm eq}(n)^{-1/2}e^{-s} - \delta_{n,n'}\lambda(n)\right]P_{\rm eq}(n)^{-1/2} \\ &= P_{\rm eq}(n')^{1/2}\tilde{L}^s_{n',n}P_{\rm eq}(n)^{-1/2}, \end{aligned} \tag{B.25}$$

where we defined

$$\tilde{L}^s_{n',n} = P_{\rm eq}(n')^{1/2}w(n' \to n)P_{\rm eq}(n)^{-1/2}e^{-s} - \delta_{n,n'}\lambda(n). \tag{B.26}$$

Here, we note that the matrix $\tilde{L}^s_{n'',n}$ is symmetric. Indeed, by using detailed balance condition, we have

$$\begin{aligned} \tilde{L}^s_{n',n} &= P_{\rm eq}(n')^{1/2}w(n' \to n)P_{\rm eq}(n)^{-1/2}e^{-s} - \delta_{n,n'}\lambda(n) \\ &= P_{\rm eq}(n')^{1/2}\left[P_{\rm eq}(n)w(n \to n')P_{\rm eq}(n')^{-1}\right]P_{\rm eq}(n)^{-1/2}e^{-s} - \delta_{n,n'}\lambda(n) \\ &= P_{\rm eq}(n)^{1/2}w(n \to n')P_{\rm eq}(n')^{-1/2}e^{-s} - \delta_{n',n}\lambda(n) \\ &= \tilde{L}^s_{n,n'}. \end{aligned} \tag{B.27}$$

Then, we recall a variational principle for symmetric matrices, which is well-known in quantum physics. By applying it to $\tilde{L}^s_{n',n}$, we have

$$E = \max_{\Phi^0 > 0} \frac{\sum_{n,n'} \Phi^0(n')\tilde{L}^s_{n',n}\Phi^0(n)}{\sum_n \Phi^0(n)^2}, \tag{B.28}$$

where E is the largest eigenvalue of $\tilde{L}^s_{n,n'}$. In this variational principle, by introducing a variational function $\Delta\widetilde{F}(n)$ by a relation

$$\Phi^0(n)^2 = P_{\rm eq}(n)e^{-\Delta\widetilde{F}(n)}, \tag{B.29}$$

we change the variational parameter from $\Phi^0(n)$ to $\Delta\widetilde{F}(n)$. The variational functional is also rewritten as follows

$$
\begin{aligned}
\frac{\sum_{n,n'} \Phi^0(n')\tilde{L}^s_{n',n}\Phi^0(n)}{\sum_n \Phi^0(n)^2} &= \frac{\sum_{n,n'} \Phi^0(n')P_{\rm eq}(n')^{-1/2}L^s_{n',n}P_{\rm eq}(n)^{1/2}\Phi^0(n)}{\sum_n \Phi^0(n)^2} \\
&= \frac{\sum_{n,n'} e^{-\Delta\widetilde{F}(n')/2}L^s_{n',n}P_{\rm eq}(n)e^{-\Delta\widetilde{F}(n)/2}}{\sum_n P_{\rm eq}(n)e^{-\Delta\widetilde{F}(n)}} \\
&= \sum_n \frac{P_{\rm eq}(n)e^{-\Delta\widetilde{F}(n)}}{\sum_{n'} P_{\rm eq}(n')e^{-\Delta\widetilde{F}(n')}} \sum_{n'} e^{-\Delta\widetilde{F}(n')/2}L^s_{n',n}e^{\Delta\widetilde{F}(n)/2}.
\end{aligned}
\tag{B.30}
$$

Thus, by introducing $\widetilde{P}(n)$ as

$$
\widetilde{P}(n) = \frac{P_{\rm eq}(n)e^{-\Delta\widetilde{F}(n)}}{\sum_{n'} P_{\rm eq}(n')e^{-\Delta\widetilde{F}(n')}},
\tag{B.31}
$$

and by using the explicit expression of the transition rate $w(n \to n')$ in the variational functional, we arrive at

$$
\frac{\sum_{n,n'} \Phi^0(n')\tilde{L}^s_{n',n}\Phi^0(n)}{\sum_n \Phi^0(n)^2} = \sum_n \tilde{P}(n)\left[\tilde{\lambda}(n) - \lambda(n)\right],
\tag{B.32}
$$

where $\widetilde{\lambda}(n)$ is defined as (3.23). This functional is exactly the same expression as the variational functional in (3.21). Therefore, by combining it with (B.28) and noticing the fact that the largest eigenvalue of $L^s_{n',n}$ and $\tilde{L}^s_{n',n}$ are the same, we obtain (3.21).

Appendix C
For Chapter 4

C.1 Derivation of (4.64)

Here, we derive (4.64). We consider a joint distribution function of $y(0)$, $y(t)$, $W(t)$ and $Q(t)$ defined as

$$P(y_0, y, W, Q, t|p) = p(y_0) \left\langle \delta(y(t) - y)\delta(W(t) - W)\delta(Q(t) - Q) \right\rangle_{y_0} . \tag{C.1}$$

By using the Langevin equations (4.6), (4.15), and (4.16), we derive the Fokker-Planck equation for $P(y_0, y, W, Q, t|p)$ as

$$\frac{\partial P}{\partial t} = \mathcal{L}_{\mathrm{FP}}^{(y,W,Q)} \cdot P \tag{C.2}$$

with the Fokker-Planck operator $\mathcal{L}_{\mathrm{FP}}^{(y,W,Q)}$ given as

$$\begin{aligned}
\mathcal{L}_{\mathrm{FP}}^{(y,W,Q)} \cdot \varphi = &-\frac{\partial}{\partial y}\left[\left(-\frac{1}{\gamma}\frac{\partial}{\partial y}U(y) + v\right)\varphi\right] - v\left(\frac{\partial}{\partial y}U(y)\right)\frac{\partial}{\partial W}\varphi \\
&-\left[\frac{1}{\gamma}\left(\frac{\partial U(y)}{\partial y}\right)^2 - \frac{T}{\gamma}\frac{\partial^2}{\partial y^2}U(y)\right]\frac{\partial}{\partial Q}\varphi + \frac{T}{\gamma}\frac{\partial^2}{\partial y^2}\varphi \\
&+\frac{T}{\gamma}\left(\frac{\partial U(y)}{\partial y}\right)^2\frac{\partial^2}{\partial Q^2}\varphi - \frac{2T}{\gamma}\frac{\partial^2}{\partial Q \partial y}\left[\frac{\partial U(y)}{\partial y}\varphi\right].
\end{aligned} \tag{C.3}$$

We multiply (C.3) by $e^{Wh_w + Qh_q}$ and integrate it with respect W and Q. Finally by noticing the definitions of the biased distribution function P_{h_w,h_q} and q_{h_w,h_q} given as (4.20) and (4.63), we obtain (4.64).

T. Nemoto, *Phenomenological Structure for the Large Deviation Principle in Time-Series Statistics*, Springer Theses,
DOI 10.1007/978-981-287-811-3

C.2 Derivation of (4.77)–(4.79) with the Cole-Hopf Transformation

Here, (4.77)–(4.79) are derived from the largest eigenvalue problems of $\mathcal{L}^{(y)}_{h_w,h_q}$ and $\mathcal{L}^{(y)\dagger}_{h_w,h_q}$. We note that Similar calculations were seen in Refs. [1, 2], for example.

We first consider the largest eigenvalue problem of $\mathcal{L}^{(y)\dagger}_{h_w,h_q}$,

$$\mathcal{L}^{(y)\dagger}_{h_w,h_q} \cdot \phi_0 = \nu_0 \phi_0. \tag{C.4}$$

We divide this equation by ϕ_0 and simplify the obtained expression. The result is

$$\begin{aligned}\nu_0 = (h_q + h_w) v \frac{\partial U}{\partial y} + \frac{T}{\gamma}\left[\frac{\partial}{\partial y}\left(\log\phi_0 - h_q U\right)\right]^2 \\ + \left(-\frac{1}{\gamma}\frac{\partial}{\partial y}U + v\right)\frac{\partial}{\partial y}\left(\log\phi_0 - h_q U\right) + \frac{T}{\gamma}\frac{\partial}{\partial y}\left[\frac{\partial}{\partial y}\left(\log\phi_0 - h_q U\right)\right].\end{aligned} \tag{C.5}$$

We then introduce a potential function $V_0(y)$ as

$$V_0(y) = -2T\left(\log\phi_0(y) - h_q U(y)\right). \tag{C.6}$$

This means we change the function we consider, $\phi_0(y)$, to the new one $V_0(y)$. This changing (or transformation) is called the Cole-Hopf transformation. We substitute this potential function (C.6) into (C.5), and we combine it with (4.61) and (4.70). This leads to

$$\mathcal{M}_{h_w+h_q,v} \cdot \left(-\frac{\partial V_0}{\partial y}\right) = 2T G_{\text{scaled}}(h_w, h_q), \tag{C.7}$$

Next, the largest eigenvalue problem of $\mathcal{L}^{(y)}_{h_w,h_q}$,

$$\mathcal{L}^{(y)}_{h_w,h_q} \cdot \psi_0 = \mu_0 \psi_0 \tag{C.8}$$

is considered. By dividing this equation by ψ_0 and simplifying it, we obtain

$$\begin{aligned}\mu_0 = -\left(\frac{1}{T} + h_q + h_w\right)(-v)\frac{\partial U}{\partial y} + \frac{T}{\gamma}\left[\frac{\partial}{\partial y}\left(\log\psi_0 + \left(h_q + \frac{1}{T}\right)U\right)\right]^2 \\ + \left(-\frac{1}{\gamma}\frac{\partial U}{\partial y} - v\right)\frac{\partial}{\partial y}\left[\log\psi_0 + \left(h_q + \frac{1}{T}\right)U\right] \\ + \frac{T}{\gamma}\frac{\partial}{\partial y}\left[\frac{\partial}{\partial y}\left(\log\psi_0 + \left(h_q + \frac{1}{T}\right)U\right)\right].\end{aligned} \tag{C.9}$$

We then define

$$\tilde{V}_0(y) = -2T\left[\log\psi_0(y) + \left(h_q + \frac{1}{T}\right)U(y)\right] \tag{C.10}$$

and substitute it into (C.9). Finally, by combining it with (4.70), we arrive at

$$\mathcal{M}_{-\beta-h_w-h_q,-v}\cdot\left(-\frac{\partial\tilde{V}_0}{\partial y}\right) = 2TG_{\text{scaled}}(h_w, h_q). \tag{C.11}$$

We note that the sign of the velocity in the left-hand side of (C.11) is reversed, which represents a reversed protocol of moving the periodic potential.

From these results (C.6), (C.7), (C.10) and (C.11), we obtain (4.77)–(4.79). Here, we mention that the uniqueness of the solution of the non-linear eigenvalue problem (4.74) is guaranteed from the Perron-Frobenius theory, since (4.74) can be rewritten as the same form as (C.4), by following the (reversed) calculation, from (C.7) to (C.4).

C.3 Derivation of (4.83) and (4.84) with Boundary Layer Analysis

Here, we derive (4.83) and (4.84), by using boundary layer analysis. First, the outer solution of (4.74), $\tilde{w}^{\mathrm{o}}_{h,v}(Y)$, is considered, which satisfies

$$\left|\frac{\partial\tilde{w}^{\mathrm{o}}_{h,v}(Y)}{\partial Y}\right| \approx \left|\tilde{w}^{\mathrm{o}}_{h,v}(Y)\right|, \tag{C.12}$$

where $\approx$ means that the left-hand side and the right-hand side are the same order of magnitude with respect to L. We then solve (4.74) for $\tilde{w}_{h,v}(Y)$, which can be easily done, because this is just a quadratic equation of $\tilde{w}_{h,v}(Y)$. The result is

$$\tilde{w}_{h,v}(Y) = \gamma\left[-v + \frac{1}{\gamma}\frac{\partial U(YL)}{\partial y} \pm \left|v(1+2Th) - \frac{1}{\gamma}\frac{\partial U(YL)}{\partial y}\right|\sqrt{1+R(Y)}\right], \tag{C.13}$$

where $R(Y)$ is defined as

$$R(Y) \equiv \frac{-v^2 4Th(1+Th) + 2K_{h,v}/\gamma}{[v(1+2Th) - (1/\gamma)\partial U(YL)/\partial y]^2} + \frac{-2T/(\gamma^2 L)\partial\tilde{w}_{h,v}/\partial Y}{[v(1+2Th) - (1/\gamma)\partial U(YL)/\partial y]^2}. \tag{C.14}$$

Now, we set $\tilde{w}_{h,v}(Y) = \tilde{w}^{\mathrm{o}}_{h,v}(Y)$ in the right-hand side of this $R(Y)$. Then, according to the following argument, we find that $R(Y)$ is negligible: By using (C.12) and (4.7), we neglect the second term of $R(Y)$. We then assume that $K_{h,v} = O(1)$, where $K_{h,v}$ is equal to $2TG_{\text{scaled}}(h_w, h_q)|_{h_w+h_q=h}$.[1] By using this assumption and (4.7), we also neglect the first term of $R(Y)$. Therefore, we omit $R(Y)$ in $\tilde{w}_{h,v}$. The result leads to an expression of the outer solution $\tilde{w}^{\text{o}}_{h,v}(Y)$ as

$$\tilde{w}^{\text{o}}_{h,v}(Y) = \begin{cases} 2T\gamma hv \\ 2\gamma\left[-v(1+Th) + (1/\gamma)\partial U(YL)/\partial y\right]. \end{cases} \tag{C.15}$$

Finally, we connect these two expressions in $\tilde{w}^{\text{o}}_{h,v}(Y)$. For this purpose, we assume the following things: Firstly, the number of the connecting points is minimized. Secondly, the one of the connecting points is located as the one where the potential has a discontinuity. Especially, in the case of $U = U_{\text{harmo}}$ and $U = U_{\text{quart}}$, the derivatives of the potentials have discontinuities at $Y = \pm 1$.[2] By combining these two assumptions with the normalization condition (4.75), we uniquely determine the solution $\tilde{w}_{h,v}(Y)$ as (4.83) and (4.84).

References

[1] T. Nemoto, S. Sasa, Phys. Rev. E **83**, 030105(R) (2011)
[2] T. Nemoto, S. Sasa, Phys. Rev. E **84**, 061113 (2011)

[1]At least, we can check that this assumption is self-consistent, because we can see that $G_{\text{scaled}} = O(1)$ as shown in (4.37), which is confirmed numerically.

[2]This assumption is also confirmed numerically.

Curriculum Vitae

École Normale Supérieure de Lyon, 15 parvis René Descartes, 69007 Lyon, France
Kyoto University, Yoshidahonmachi, Sakyo Ward, Kyoto, 6068501, JAPAN
e-mail: takahiro.nemoto@ens-lyon.fr

Appointments

- JSPS Research Fellowship for Young Scientists (DC1), (April, 2012–March, 2015)
- Post-doctoral research fellow with Prof. Freddy BOUCHET, École Normale Supérieure de Lyon, (April, 2015 –).

Education

- Doctor of Science (Ph.D.) in Division of Physics and Astronomy, Graduate School of Science, Kyoto University, (2012–2015), supervisor: Shin-ichi Sasa
- Master of Arts (MA) in Group of Condensed Matter theory and Statistical Mechanics (C group), Graduate School of Arts and Sciences, The University of Tokyo, (2010–2012), supervisor: Shin-ichi Sasa
- Bachelor of Science in Department of Physics, Tokyo Institute of Technology (2007–2010).

Research Statement

A realisability of rare-event samplings in real experiments had not been explored yet. If such a sampling method existed, it certainly would lead to many applications in several fields. My work was the first positive result about it. I proposed a new rare-event sampling method to calculate large deviation functions of time-averaged quantities. The method was based on thermodynamics, so that it could potentially be used in real experiments.

Major Honors and Awards

- 2015 Young Scientist Award of the Physical Society of Japan.

T. Nemoto, *Phenomenological Structure for the Large Deviation Principle in Time-Series Statistics*, Springer Theses,
DOI 10.1007/978-981-287-811-3

MIX
Papier aus verantwortungsvollen Quellen
Paper from responsible sources
FSC® C105338

If you have any concerns about our products,
you can contact us on
ProductSafety@springernature.com

In case Publisher is established outside the EU,
the EU authorized representative is:
Springer Nature Customer Service Center GmbH
Europaplatz 3, 69115 Heidelberg, Germany

Printed by Libri Plureos GmbH
in Hamburg, Germany